T0230460

Research Techniques for Computer Science,
Information Systems and Cybersecurity

Uche M. Mbanaso • Lucienne Abrahams
Kennedy Chinedu Okafor

Research Techniques
for Computer Science,
Information Systems
and Cybersecurity

 Springer

Uche M. Mbanaso
Centre for Cybersecurity Studies
Nasarawa State University
Keffi, Nigeria

Lucienne Abrahams
LINK Centre
University of the Witwatersrand
Johannesburg, South Africa

Kennedy Chinedu Okafor
Department of Mechatronics Engineering
Federal University of Technology
Owerri, Nigeria

ISBN 978-3-031-30033-2   ISBN 978-3-031-30031-8   (eBook)
https://doi.org/10.1007/978-3-031-30031-8

This Springer imprint is published by the registered company Springer Nature Switzerland AG
The registered company address is: Gewerbestrasse 11, 6330 Cham, Switzerland

Foreword

Research and development currently plays a great role in the economic growth of nations. Particularly, research in the three fields of computer science (CS), information systems (IS), and cybersecurity (CY) is providing transformational solutions to business processes and human needs. Conducting contemporary research studies in CS, IS, and CY is often a major challenge for young academics. This book provides a very clear and structured explanation on the various types of research, including quantitative and qualitative research; the various stages of the research process; research design; research ethics; and how to avoid plagiarism. The authors clearly explain the computer science, information systems, and cybersecurity fields of study, the subtle differences and overlaps, in an interesting way, using diagrams. Since the research settings of these three related fields are unique, young researchers need to understand the distinction, as well as how they interplay with each other, for the best research outcomes for solving human, business, and organisational problems.

The ever-increasing demand for digital solutions demands multi- and interdisciplinary research approaches and innovation. This book presents the use of the funnel strategy to handle the rigorous processes of sorting out the myriad of ideas from literature, conceptualising research ideas, identifying the study goals, framing appropriate questions, scoping, and considering whether the problem is researchable. Thus, the funnel framework helps to focus the researcher's thought process on arriving at meaningful research topics and processes that lay the foundation for valuable research results.

Furthermore, originality and innovation in research emerge by design, not by accident, a big challenge for young researchers. Contemporary research requires a well-structured sequence of logical and structured thinking to drive an acceptable scientific or social enquiry and useful outcome.

One of the things that make this book unique and a 'must read' is the use of the mind mapping technique. This can be applied to research in all fields of endeavour, medicine, engineering, and related sciences. The use of mind mapping has been successfully applied in project management to simplify complex ideas, ignite creativity, and boost productivity.

The combined use of the funnel strategy and mind mapping will greatly facilitate practical research for solving evolving societal problems. The appropriate use of the funnel strategy and mind mapping will guarantee originality and innovation by design from the beginning of the research project. This book will be very useful for postgraduate students and young researchers in computational sciences and computational engineering.

Personally, I found it a very good textbook for introducing research skills in the fields of computer science (CS), information systems (IS), and cybersecurity (CY). All the chapters are self-explanatory and form a comprehensive tutorial guide for the full life cycle of any research process. As such, the book is a valuable resource that students can consult throughout their research studies and professional life, particularly in engineering and sciences. Therefore, I strongly recommend this book to postgraduate students and lecturers who are desirous of fruitful innovative research outcomes.

Dean, School of ICT   Gloria A. Chukwudebe
Federal University of Technology,
Owerri, Nigeria

Foreword

Indisputably, advances in science, and in particular, information and communications technology (ICT), have continued to alter the means and ways human beings live, interact, and associate. The evolution and revolution of the digital age is a full-blown outcome of research and development (R&D). The entire world is continually faced with evolving uncertainties as a result of the growing human development in the context of the universe we live. To surmount the growing challenges in ICT, research in the domain of computer science (CS), information systems (IS), and cybersecurity (CY) is providing unprecedented wave of advancements to enhance solutions to business processes and human enterprise. I appreciate the fact that carrying out research works in CS, IS, and CY is faced with major challenges of complexity in the field for young academics. This book offers a unique clear and concise explanation of the universe of research that takes into cognizes of diverse types of research, i.e., quantitative, qualitative, and pragmatic research. It clearly demonstrates the numerous stages of the research activities including research philosophy; research design; research ethics; data collection, presentation and analysis; and how to avoid plagiarism. This book is unique in the sense that the authors brought in one place the computer science, information systems, and cybersecurity fields of study together, the blurring lines, and intersections, in a stimulating manner, using practical examples. Notably, the research activities of these three related domains are inimitable; new researchers need to appreciate the differences, as well as how they are related with each other, to achieve better research results that address the universe of human issues, business enterprise, and organisational problems.

There is also a question of growing demand for digital transformations, which requires a diverse approach in a multi- and interdisciplinary field such as CS, IS, and CY. Interestingly, this book introduces a novel funnel strategy, which can help young researcher grapple with research inquiry, a unique way to start the journey into research activities. The funnel strategy offers an exceptional way to address the rigorous activities of sorting out the initial ideas from finding a topic of interest; scoping the literature, conceptualising fresh research ideas, constructing the study goals, constructing appropriate questions, and determining whether the problem is researchable. Consequently, the funnel context can help a researcher to remain focused on the nuances of thought process to reaching a meaningful subject area and activities that give the foundation for appreciable research outcomes.

Another giant stride in this book is the introduction of mind mapping techniques, which can be applied to all fields of endeavour, social science, environment science, engineering, and other sciences in conceptualising research concepts and building a sort of knowledge tree that can deepen the understanding of the phenomenon under inquiry. The application of mind mapping in project management is popular; however, this is the first time it is formally introduced to streamline complexity in research ideas, kindle innovation and creativity, as well as improve productivity.

Undisputedly, this unique book is a 'must read' for postgraduate students, and in particular PhD researchers pursuing research in CS, IS, and CY.

I strongly recommend this textbook to postgraduate students as it pragmatically introduced research activities and processes in the fields of computer science (CS), information systems (IS), and cybersecurity (CY) in a phased-out fashion. The chapters are presented in chorological order, making the book self-explanatory and a holistic tutorial masterpiece for understanding the full life cycle of any research endeavour. I am persuaded that this shall remain a valuable handbook for scholars who wish to maintain competative edge as top-rated researchers, particularly in CS, IS, and CY. I commend the authors, who have worked so hard to produce this scientific book, and encourage them to constantly update the book to match the pace of advancement in these fields.

The Vice Chancellor   Suleiman Bala Mohammed
Nasarawa State University
Nigeria

Comments from Reviewers

The computing sciences are founded upon principles and theories from various science and engineering fields, and hence draw inspiration and adopt techniques from these areas. While conducting research in the CS, IS, and CY fields, the researcher needs to apply a number of separate and independent skills from many other areas and must be open minded and flexible in formulating the design and conducting the entire research process. Most books on conducting research cover a broad spectrum of research fields and rarely narrow down on specific subject areas, thereby diverting the attention of novice researchers and students from focusing on their particular areas of research. In this book, the authors make good use of their broad research skills and considerable knowledge of digital technology to present a pragmatic approach that guides researchers on conducting contemporary research in the field of computing sciences. However, they still cover the basic review of the general skills and methods required to undertake research in a wide range of subjects, thereby benefiting researchers in all fields. With its numerous real-life examples and illustrations, the book is a perfect resource and a practical guide to researchers and students.

Professor N. V. Blamah
Professor of Computer Science
Department of Computer Science
University of Jos, Jos, Nigeria

One of the things that makes this book unique, and a 'must read' is the use of the mind mapping technique. This can be applied to research in all fields of endeavour, medicine, engineering, and related sciences. The use of mind mapping has been successfully applied in project management to simplify complex ideas, ignite creativity, and boost productivity. The combined use of the funnel strategy and mind mapping will greatly facilitate practical research for solving evolving societal problems. The appropriate use of the funnel strategy and mind mapping will guarantee originality and innovation by design from the beginning of the research project. This book will be very useful for postgraduate students and young researchers in computational sciences and computational engineering.

Professor Gloria A. Chukwudebe
Professor of Computing and Electronics Engineering
Federal University of Technology, Owerri, Nigeria

Research in the field that broadly goes under the banner of 'information and communication technology', or ICT, is by its very nature cross-disciplinary. For those carrying out research in areas of ICT, and for those supervising and reviewing such work, a challenge always arises as to which disciplinary silo the work belongs. Is it engineering, or science, or business science, or philosophy? Each of these individual disciplines has its own tradition of research methods and practice. ICT research invariably cuts across these traditions. Researchers are often forced to fit their research into a limited and inappropriate disciplinary box. This book makes a significant contribution by setting out a coherent framework for ICT research acknowledging its inherently broad and cross-disciplinary nature.

Emeritus Professor Barry Dwolatzky
Director of Innovation Strategy, Office of the Deputy Vice-Chancellor: Research &
*   Innovation, University of the Witwatersrand, Johannesburg*
Director, Joburg Centre for Software Engineering (JCSE), School of Electrical and
*   Information Engineering, University of the Witwatersrand, Johannesburg*

Having been in computing for more than two decades in teaching and learning, research, knowledge development and most recently in academic mentoring, I can confidently say that one changing and diverse factor in the computing industry, especially in the academic arena, is having a specifically defined method of research. Academics in different fields of computing tend to conduct their research using different approaches because most of the problems that involve computer solutions are very diverse and hence using a single method or approach becomes inappropriate. In elucidating knowledge and trying to provide a balance between these differing fields that demand computing solutions, this book is handy and good company on this journey of uncertainties and unending diversities. I recommend that this book becomes a handbook for all students and staff in computing fields in universities, especially in developing economies. It will help in providing a roadmap for conducting research that meets international standards, as well as enhancing personal development and knowledge creation.

Dr Elochukwu Ukwandu
Lecturer in Computer Security and Fellow of Higher Education Academy,
Cardiff Metropolitan University, UK

The book is well-researched. The authors' approach in presenting and discussing contemporary research in computing is quite lucid and easy to understand. Chapter 1 gives a good introduction to research. One of the challenges we face, particularly at the PhD level (computer science), is that students tend to avoid engaging in problems that are challenging – unfortunately, that is actually what research at the PhD level in computer science is about. Basic research involves developing and testing theories and hypotheses that are intellectually challenging. The knowledge produced through basic research is sought in order to add to the existing body of knowledge. Simply put, the research student should understand that basic research is a tool for the development of a solid foundation for reliable knowledge. The prominence given to critical thinking in Chap. 4 is well-deserved. Overall, the book is very impressive.

Professor Nwojo Agwu Nnanna
Professor of Computing and Mathematics
Department of Computer Science, Nile University of Nigeria

Research is foundational to the advancement of technology and civilisation. Many postgraduate researchers embark on their projects with ample enthusiasm and technical skill, but often lack training in the practical aspects of research necessary to conduct their projects and realise their ambitions. The questions faced by new researchers can be daunting: 'How do I select appropriate research questions for a project proposal? What is the best way to conduct a literature review? What types of data will I need to collect to answer my questions?' This book addresses these questions and provides a broad coverage of contemporary research skills that will help new researchers plan and manage their projects, from initiation through to completion.

Dr Barry Bentley
Senior Lecturer and Head, Bioengineering Research Group,
Deputy Director, EUREKA Robotics Centre,
Cardiff School of Technologies,
Cardiff Metropolitan University,
Cardiff, Wales, United Kingdom

For the most part the book is a general guide to academic research covering both the scientific and social science research designs, particularly where interdisciplinary and multidisciplinary research is required. It is timely as computing research is increasingly both interdisciplinary and multidisciplinary due to rapid digital transformation of many sectors that we are witnessing, leading to emergence of new sub-fields such as agritech, fintech, industry 4.0 to name a few. The research in these is intricately linked to the three areas of research that are focus of this book. This book is very useful to postgraduate students in these new sub-fields that could be located outside the traditional faculties of natural sciences, engineering and business. What could be covered, in a second edition, is the link between the choice of research methodology and research design with the faculty that the research is based in. For example, where these fields are in science faculties and engineering faculties, design and creation type research methods typically dominate the research designs. This research design is less dominant where the research is located in other faculties such as in the business faculty, but still applicable.

Professor Ntsibane Ntlatlapa
Head of the C4IRSA, Council for Scientific and Industrial Research
and Honorary Adjunct Professor, LINK Centre,
University of the Witwatersrand, Johannesburg

Preface

The art and practice of impact-driven research requires creative design, effective execution and continuous revision, until the point of submission. Postgraduate research must aim to introduce new concepts and techniques to simplify the complexities associated with the computer science (CS), information systems (IS) and cybersecurity (CY) research domains.

One of the main objectives of this book is to furnish readers with a foundation that will facilitate active quantitative, qualitative and mixed methods research. The content offers important perspectives on how to think about deepening research in CS, IS and CY, noting that these subjects can be studied from a mathematical, engineering, health sciences, social sciences or interdisciplinary sciences perspective.

The book is organized into eight chapters. Chapter 1 introduces twenty-first century postgraduate research, including the concepts of research, research types, research attributes, research cycle, philosophical research design and quality in research writing, among others. Chapter 2 gives an overview of computer science (CS), information systems (IS) and cybersecurity (CY) research, and the intersection amongst them is highlighted. Chapter 3 focuses on designing the research proposal, or interim report. In this regard, the discussion explains the need to understand the research proposal as a structured motivation to undertake and produce original research. Chapter 4 presents the concept of the funnel strategy and mind mapping to visualize research design. Chapter 5 focuses on the necessary skills for research writing, and for writing the background discussion, the literature review and analytical framework. Chapter 6 sets out key issues in research theory: philosophy, design and methodology. Chapter 7 explains processes in data collection, data presentation and data analysis. Chapter 8 wraps up the book with ideas about research planning and research management.

Keffi, Nigeria
Johannesburg, South Africa
Owerri, Nigeria

Uche M. Mbanaso
Lucienne Abrahams
Kennedy Chinedu Okafor

Chapters at a Glance

Chapter 1: Twenty-First Century Postgraduate Research

The discovery and advancement of the universe of knowledge require students and academics to conduct research in ways that promote the most effective outcomes. Research enables us to probe uncertainties and uncover gaps in knowledge, in order to gain greater insight into, and understanding of, the phenomena that require investigation. In retrospect, we see that research has helped to advance the human enterprise. Through deliberate and conscious investigations and experimentations, discovery and interpretation of evidence have constantly probed human existence, generating theories or laws in the bid to establish new sets of ideas, or real-world application of such new ideas, and/or revision of existing theories or principles. This systematic inquiry that aims to discover new things and create new facts or knowledge spans all facets of life.

Chapter 2: Computer Science (CS), Information Systems (IS) and Cybersecurity(CY) Research

In this chapter, we dissect and situate the CS, IS and CY research disciplines, to assist new researchers to understand how to focus their research efforts. In contemporary research, the line between computer science (CS) research, information systems (IS), and cybersecurity research studies is not completely blurred. There are subtle differences, intersections and overlaps. While their research settings may be unique, new researchers need to understand the distinction between CS, IS and CY research enquiries, as well as how they interplay with each other. Traditionally, CS research leans more towards scientific investigations using complex mathematical theories and proofs, to solve problems or to provide answers to questions (Hevner et al., 2004). IS research, on the other hand, tends towards an application of CS to solving real-world problems concerning business, government or society.

Cybersecurity study is concerned with new knowledge in improving and/or developing products relevant to digital security, safety and privacy.

Chapter 3: Designing the Research Proposal or Interim Report

The research proposal is an important milestone in the overall process of postgraduate research. Various schools may call this document by different names, noting that it may also be called the interim report in your institution. Some institutions have a short and a long proposal. Whichever approach your institution uses, the purpose of these documents is similar. We will deal mainly with the research proposal, and we will also refer to the interim report, which is a little more developed than the proposal.

Chapter 4: Adopting a Funnel Strategy and Using Mind Mapping to Visualize the Research Design

Academic research is more than a project or a piece of writing. A research paper, dissertation or thesis can only be written up after relevant research has been conducted. Research students will have different motivations for conducting a research inquiry, some will have a specific problem they would like to address through experimentation, while others will wish to add to theory. Designing a research study is often a major challenge for novice researchers due to a limited understanding of the research process, and possibly also the particular meaning of research (Carlsson, 2006) in the context of a specific research problem. Many academic researchers are uncertain how to identify and craft the research problem to be addressed in the inquiry and how to frame the main research question. The beginner must conceptualize a research idea, set the scope of the research problem and consider whether the problem is researchable (Pajares, 2007). However, finding a researchable area may not be straightforward. This is where the funnel strategy is useful. The funnel strategy is a mental model, which a researcher can adapt, to provide guidance on the research process, enabling the flow of thoughts and ideas throughout the research journey.

Chapter 5: Foundational Research Writing, Background Discussion and Literature Review

Before getting to the practical task of data collection, it is important to evaluate the foundational components for undertaking the research endeavour, namely the background discussion and the literature review. In this chapter, we discuss their content and their respective roles and significance in the CS, IS and CY research study. We need to know how foundational research writing, the background discussion section and the literature review fit into the full range of research activities and how they enable the researcher to complete the study. Foundational research skills include vital processes, from framing a statement of the research problem to designing the analytical framework.

Chapter 6: Research Philosophy, Design and Methodology

Research philosophy, design and methodology are critical components of scholarly research; hence, it is important to examine each of these components carefully in the preparation of CS, IS and CY research studies. We need to understand where and how each of these elements fits into the research process. Ideas about research philosophy, research design and research methodology relate to the conceptual thinking about what constitutes new knowledge and how to produce new knowledge, as well as the ethics applied when engaging in research. A researcher's philosophical choices shape the assumptions, perceptions and interpretation of the nature of reality. Our individual research philosophy in relation to a particular research project influences how we see the research problem, how we understand data, what analysis we make, what conclusions we draw and what theories we advance. For example, in some studies we may take a human-centric approach, in other studies we may take a techno-centric approach and on occasion we may balance the two in a techno-human approach. Consequently, a researcher must be grounded in understanding the need to adopt a clearly stated research philosophy, early in the research process. Moreover, these three elements form the structural basis that examiners use to evaluate and validate a research study. Typically, an inappropriate research philosophy, design and methodology may result in invalid study outcomes, where the interpretations and conclusions are faulty. In this chapter, the rudiments of research philosophy, design and methodology are explained within the domains of CS, IS and CY. Babbie (2010) is a good companion text to this chapter.

Chapter 7: Data Collection, Presentation and Analysis

Some of the most important components of postgraduate research are data collection, data analysis and data presentation, because these are the foundations that the 'new knowledge' stands on. These components establish the basis on which reviewers, examiners and readers evaluate the final research report, dissertation or thesis. In this chapter, we discuss data collection, data analysis and presentation, relevant to studies in computer science, information systems and cybersecurity. Data collection and data analysis techniques differ with respect to quantitative and qualitative research and we will discuss some of the key characteristics of each approach in this chapter.

Chapter 8: Research Management: Starting, Completing and Submitting the Final Research Report, Dissertation or Thesis

Managing the research process is a vital contributor to successful completion. Research involves a precise set of tasks and activities. This chapter focuses on the practical experiences and good practices garnered by the authors when supervising postgraduate research researchers. The chapter addresses research planning, the supervisory relationship, resource management, time management as well as domestic issues that usually impinge on research time. The goal of research management is to complete the research process and to present the completed research report, dissertation or thesis for examination.

Acknowledgements

A book of this nature has had strong intellectual inputs from diverse experts in the field of CS, IS and CY. We are grateful to all those who reviewed and evaluated this edition. In particular, we are indebted to Professor N. V. Blamah, whose contributions helped to shape this first edition; Professor Gloria A. Chukwudebe, who reviewed the manuscript and wrote the foreword for this book; Dr Barry Bentley, for taking ample time to provide useful feedback; Professor Nwogo Agwu Nnanna, for providing feedback on the need to demonstrate how some of the research concepts introduced in this edition can be applied in research; Dr Elochukwu Ukwandu, for providing useful feedback that helped improve the quality of the book; and to Professor Dwolatzky and Professor Ntlatlapa for their review comments. We thank Mr Julius Ato for structuring and formatting the manuscripts. We appreciate all our Master's and PhD candidates who provided valuable feedback about the readability of this edition.

Finally, we thank our various families, from who we borrowed most of their time to complete this book.

Contents

List of Figures

List of Tables

Acronyms[1]

AI	Artificial intelligence
ANN	Artificial neural network
APA	American Psychological Association
ATC	Air traffic controller
CS	Computer science
CRMM	Cybersecurity resilience maturity measurement
CY	Cybersecurity
ERP	Enterprise resource planning
Gbps	Gigabytes per second
GDI	Global diffusion of the Internet
GDP	Gross domestic product
HCI	Human-computer interaction
ICT	Information and communication technology
IEEE	Institute of Electrical and Electronics Engineers
IoT	Internet of Things
IS	Information systems
IT/IS	Information technology/Information system
ITU	International Telecommunication Union
ISACA	Information Systems Audit and Control Association
MLA	Modern Language Association of America
NREN	National research and education networks
PhD	Doctor of philosophy
PII	Personally identifiable information
QDA	Qualitative data analysis
R&D	Research and development
SDLC	Software development life cycle
SEK	SEMAT Essence Kernel

---

[1] The following acronyms are used in the text.

SLR	Systematic literature review
SMART	Specific, measurable, achievable, relevant, timely
SWOT	Strengths, weaknesses, opportunities, threats
TURF	Total unduplicated reach and frequency analysis
WEF	World Economic Forum

About the Authors

Uche M. Mbanaso, PhD is a leading cybersecurity subject matter expert. He is currently the Executive Director, Centre for Cyberspace Studies, and an Associate Professor in Cybersecurity and Computer Science at Nasarawa State University, Keffi, Nigeria. He is also a visiting academic at the LINK Centre, University of the Witwatersrand, Johannesburg, South Africa. He is a key player in the development of cybersecurity in Nigeria, having played important roles in the development of cybersecurity policy and strategy, data and privacy protection, and many other cybersecurity initiatives in Nigeria. He is a Software Security Architect, and writes software in Java, C# and Python. He has strong competence in applied cryptography, especially in distributed environments where digital trust is a requirement. Prior to joining Nasarawa State University, he worked with the public sector for over 25 years during which he actively functioned in providing cutting-edge secure ICT solutions. His research and scholarly work include a TETFund sponsored research project on Cybersecurity and Critical National Infrastructure (CNI). He has published over 25 researched journal articles and 30 conference proceedings in the areas of cybersecurity and information and communications security. He possesses the aptitude for mentoring computing and cybersecurity mentees, especially undergraduate and graduate students. Prof Mbanaso supervises PhD candidates applying well-developed pedagogical skills. He is actively mentoring young students in tech skilling-up programmes for dynamic cybersecurity, embedded programming and

robotics. Prof Mbanaso earned an undergraduate quali-
fication in Electronics and Communications
Engineering in Nigeria, MSc in Information
Technology in the UK and PhD in Communications
and Information Security in the UK. He is a member of
ISACA, IEEE and Nigeria Computer Society (NCS)
and an editorial board member of *The African Journal
of Information and Communication*, https://ajic.wits.
ac.za/ indexed in the SciELO Citation Index.

Lucienne (Luci) Abrahams, PhD is Director of the
LINK Centre at the University of the Witwatersrand,
Johannesburg   https://www.wits.ac.za/linkcentre/,
building research on digital innovation and how digital
technologies and processes influence change. Studies
include: case studies in digital governance; digital
skills gap analysis; digital strategy; scaling up innovation in tech hubs; open access in scholarly publishing
(Open AIR research partnership) and health e-services
improvement (Egypt-South Africa research partnership). Luci convenes the MA and PhD programmes in
Interdisciplinary Digital Knowledge Economy Studies
and supervises postgraduate research; lectures on short
courses on disruptive technologies, digital operations
and leadership; and lectures on research methods for
cybersecurity professional practice. She serves on the
Board of the Tertiary Education Network (TENET).
Luci is Corresponding Editor for *The African Journal
of Information and Communication*, https://ajic.wits.
ac.za/ indexed in the SciELO Citation Index. She has
previously served on many Boards and Committees,
including the Development Bank of Southern Africa,
the National Advisory Council on Innovation, the
Board of the National Research Foundation, the
Financial and Fiscal Commission, the Ministerial
Review Panel on the Science Technology and
Innovation Institutional Landscape and on the
Evaluation and Review Reference Group for the review
of the National Research Foundation. She chaired the
Advisory Committee on ICT Governance for the
University of Fort Hare (2019–2020). She designs and
presents workshops on digital universities, digital
skills and future oriented policy and regulation for the
digital era. In this book, she shares her knowledge
gained from supervising postgraduate students.

Kennedy Chinedu Okafor, PhD is a Senior Member, IEEE, USA; Chair of IEEE Consultants Network AG-Nigeria; and a Senior Teaching Researcher with the Department of Mechatronics Engineering, Federal University of Technology, Owerri, Nigeria. Kennedy is a World Bank Faculty Member at the Africa Centre of Excellence for Sustainable Power and Energy Development (ACESPED), University of Nigeria Nsukka (UNN), Nigeria. In 2017, Kennedy received the prestigious Vice-Chancellor Award as the overall best graduating PhD candidate from the Faculty of Engineering, UNN. He holds a research Fellowship appointment with Imperial College London and Manchester Metropolitan University (SIIRG) as a Postdoctoral Researcher in Driverless Car Technology. Also, Kennedy is a Senior Research Associate with the University of Johannesburg, South Africa (2021–2022). He was formerly an industry collaborator with the Center for Lighting Enabled Systems & Applications (LESA), USA. He formerly worked for the National Agency for Science and Engineering Infrastructure (NASENI), Nigeria, as a Senior R&D Engineer. Kennedy has certifications from IEEE, Cisco, IBM and CompTIA A+®. Kennedy has organized over 50 IEEE events while serving as a keynote speaker in Africa, the Middle East and Europe. He is an author of 3 university books, 2 monographs, 5 book chapters and 150+ publications (indexed in Google Scholar, Scopus, and Web of Science, SCIE | Publons). He has reviewed over 300 papers and handled over 320 editorial tasks (both in Scopus and Web of Science). Kennedy has supervised/co-supervised and mentored several MSc and PhD students. He serves on the editorial boards of IEEE, Elsevier, Springer, Taylor & Francis, Wiley and IGI Global Journals. Kennedy is currently collaborating with Rensselaer Polytechnic Institute USA, IBM, Cisco Systems USA, National Instruments USA, Smart Power, and Energy Systems Research Group (SPESRG) University of Johannesburg, South Africa. He is the Executive Secretary of the Association of Artificial Intelligence and Machine Learning of Nigeria (AAIMLON). He is a Certified COREN licensed engineer and a Fellow of African Scientific Institute, USA.

Chapter 1
Twenty-First Century Postgraduate Research

1.1 Introduction

The discovery and advancement of the universe of knowledge requires students and academics to conduct research in ways that promote the most effective outcomes. Research enables us to probe uncertainties and uncover gaps in knowledge, in order to gain greater insight into, and understanding of, the phenomena that require investigation. In retrospect, we see that research has helped to advance the human enterprise. Through deliberate and conscious investigations and experimentations, discovery and interpretation of evidence have constantly probed human existence, generating theories or laws in the bid to establish new sets of ideas, or real-world application of such new ideas, and/or revision of existing theories or principles. This systematic inquiry that aims to discover new things and create new facts or knowledge spans all facets of life.

Research is a purposeful and organized endeavour pursued to highlight new or improved knowledge of things. This knowledge could be found in nature, in society, in cultural diversity or even in abstract phenomena. Furthermore, research can be the basis to validate or establish assumptions, restate the findings of existing studies or address fresh or prevailing questions of the day. It can be the premise for sustenance the development of propositions or develop new theories or laws. Research requires careful work, attention to detail and focus, implying a thoughtful inquiry based on established methods to aid in discovery through investigations, experiments and clarification of actualities. It can further extend to the application of such new findings, theories or laws in solving real-world problems in advancing the human enterprise. Consequently, research may involve the collection of data and analysis and interpretation about a particular phenomenon or events. Thus, researching takes some sort of rigorous process that is systematic and structured aiming to find new ideas, understanding and knowledge to advance human inventiveness.

U. M. Mbanaso et al., *Research Techniques for Computer Science, Information Systems and Cybersecurity*, https://doi.org/10.1007/978-3-031-30031-8_1

Research can equally be referred to as an organized activity that has plan and structure, steps, phases and a philosophical way of thinking (or critical thinking), to observe, gather and analyse information that can expose new findings, or establish evidence, as the basis for the advancement of knowledge. Invariably, conducting research means a cautious and thorough investigation of a specific issue or an event, concern or problem, leveraging empirical methods and techniques. According to Oates (2006), an investigation or inquiry qualifies as research, if it satisfies 'the six Ps i.e., purpose, paradigm, process, participants, products, and presentation'.

Every research activity must be founded on underlying philosophical assumptions, which can help define the concept of 'valid' research, based on the methodological stance. The methodological approaches accentuate the development of the novel knowledge in a given area of study. In particular, research in the broad arena of digital transformation, and in the specific fields of computer science (CS), information systems (IS) and cybersecurity (CY) (Jungmann et al., 2015; Raunio et al., 2019), as well as in the strongly emerging field of cybersecurity, is witnessing increasing methodological diversity. This may be because contemporary research on the many facets of digital transformation revolves around organization, people and technology (Javaid & Iqbal, 2017) in their interconnection, which itself is changing, needing new approaches to study. This increasing openness buttresses the application of innovative research methodologies in both disciplines, as well as in more complex disciplines like mechatronics, and other converging scientific fields.

Conducting contemporary CS, IS and CY research requires the understanding of many facets of investigative philosophy. CS, IS and CY disciplines are concerned with the development of computing theories and principles and the application of computing (or information system) to solve human, business and organizational problems. Arguably, many CS, IS and CY students grapple with how to begin the research journey, how to differentiate a developmental project from a research project and how to design the research study. This book aims to guide research beginners in conducting empirical and scientific research inquiry. Notwithstanding that the book focuses mainly on CS, IS and CY, many of the chapters may be useful to researchers in related fields, for example, the applications of software to language usage.

1.2   The Concept of Research

1.2.1   Scientific Paradigm

Research refers to a scientific, in other words, systematic way of investigating a problem, in order to obtain a deep understanding of the research problem and in order to expand the frontiers of knowledge with respect to that problem and the field more broadly (Mulhanga et al., 2014), for the benefit of society. In some cases, there may not be any immediate plan to use the research findings or the knowledge created. In practice, research can be summarized as a purposeful search for an answer to a specific question, which needs an answer. Simply put, research typically originates from a question or known problem, although an invention may not necessarily be the outcome of such research.

1.2.2   Scientific Philosophy

To most students in the CS, IS and CY domains, the concept of research seems complex. CS, IS or CY research is conducted when it involves a well-organized method of gathering and analysing data and interpreting and presenting findings in order to improve and advance CS, IS and CY in solving organizational and business problems. At the centre of every research endeavour is the creation of new ideas, and this encompasses carefully structured and rigorous activities aiming to bring modified or new evidence (data), theories or frameworks. With the advent of the Internet, CS, IS and CY research is growing in the areas of agricultural production, business, development of cities, education, health services, manufacturing and mining, indeed in almost all economic sectors and social dimensions. Thus, there is a rising demand for academic, industry and freelance researchers to actively engage in finding new and innovative ideas. The quest for new ideas is prompting stimulation of critical thinking on how to match human problems with real-world technology-driven digital solutions.

Therefore, CS, IS and CY research must involve a known or verifiable scientific philosophy and process. This is very important for finding a proven solution/answer to a specific problem, using methods that are repeatable. It implies that conducting CS, IS and CY research must use a variety of scientific methodologies to explore, experiment, analyse and report research outcomes, including quantitative, qualitative and mixed-methods methodologies. We will discuss this further in a later chapter.

1.2.3   Ethics and Avoiding Plagiarism

Researchers must be concerned with three main considerations, namely, plagiarism, copyright and ethics in scholarly production, all of which are necessary for the responsible conduct of research. Plagiarism refers to the 'theft of ideas, misappropriation of intellectual property and the substantial unattributed textual copying of another's work' (Roig, 2013). Many universities have plagiarism explanations and relevant explanatory documents publicly available on their websites for students to consult. To better understand plagiarism, you can read the available documents on the website of the University of the Witwatersrand, on the page Plagiarism, Citation and Referencing Styles: Wits Plagiarism Policy; see various guidelines available at

https://libguides.wits.ac.za/plagiarism_citation_and_referencing/Avoidance
https://libguides.wits.ac.za/plagiarism_citation_and_referencing/plagiarism_policy

1.3   Research Types

There are two broad types of research (Nnabude et al., 2009), which depend on varying perspectives and backgrounds, and are explained as follows:

1.3.1   Basic/Fundamental/Theoretical Research

This type of research is carried out purely for the acquisition of knowledge with or without any intended application of the results. The results may not have immediate feasible commercial value or industrial relevance. However, in the cause of time, the findings may eventually become relevant and usable.

1.3.2   Applied/Practice-Oriented/Trade-Oriented Research

This type of research is carried out mainly for providing answers to real-world problems or to address developmental interests. In some instances, it can be linked to directed research and development (R&D) efforts. This implies that the findings of such research may be expected to be applied in the production of goods and services for:

- Increasing output (capacity, yield, etc.)
- Improving quality (product, lives, produce, etc.)
- Improving function (mechanisms, functionalities, etc.)
- Cost reduction (systems, processes, deployments, etc.)

With respect to these two types, scientific research sets out to achieve just one thing, that is, to discover something new, which is not trivial. This alludes to the fact that research itself is a rigorous activity, with increasing difficulty and complexity. Finding something new – new theories, principles, procedures, methods and innovative instrumentation – is a research imperative. It is worthy to note that there can never be an end to something new. This is because of the complex nature of the world and its creative models that are usually beyond our initial imagination.

1.4   Research Attributes

In general, research engagement in CS, IS and CY must have the following attributes, as highlighted by Jannach et al. (2012):

Attribute 1: Research design should be based on the established philosophical stances that are scientifically valid relative to the field of study. Where borrowing or learning from another field of study, a clear justification must be provided.

Attribute 2: The design should employ research techniques and methodological approaches that have proven dependable and reliable and that are directly relevant to the nature of the research problem.

Attribute 3: Research design should be structured to avoid unscientific hypothesis formulation, subjectivism and bias.

Attribute 4: The analytical discussion should avoid speculation and present well-reasoned analytical arguments.

Attribute 5: The research should be monitored and take place in a controlled environment (quantitative research) or within a relatively limited scope (quantitative and qualitative research).

Attribute 6: The research should adopt rigorous and structured procedures.

Attribute 7: Research should be logical and systematic.

Attribute 8: The final research output and the research manuscript should have a valid, verifiable conclusion.

Attribute 9: In some cases, the research manuscript should present an empirical conclusion.

1.5   Qualities of Research

The outcome of research depends on the attributes stated in Sect. 1.4. The implication is that these attributes are vital to turning out CS, IS and CY research of good quality, which usually will lead to:

Outcome 1: A reasonably thorough review and critical assessment of existing works

Outcome 2: Significance for a particular audience, or knowledge community, or practitioner community

Outcome 3: Technical soundness

Outcome 4: Contribution to the body of knowledge through originality and novelty

1.6   Research Cycle

Researchers should understand that a typical research cycle is an unending process of discovery and knowledge creation, as shown in the example for a quantitative research study in Figure 1.1. Such discovery and knowledge creation is the outcome of a laborious, painstaking process of structured activities. Research is a systematic expedition for probing into the world of undiscovered or unknown knowledge. The researcher should embrace this research production cycle, ensuring good planning

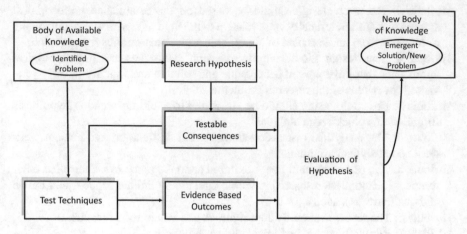

Figure 1.1 *A Generic Research Cycle for Quantitative Research*

and organization of each step, including research design, data collection, data analysis, generating conclusions and theory building or practical design.

1.7   What Is Not Academic Research?

Not all academic projects can qualify as research. In CS, IS and CY, projects carried out may not necessarily qualify to be classified as a research work, if the research lacks the attributes of research design discussed above. Some examples of activities that would not normally be classified as research include:

- Materials testing (as is the case in engineering or sciences)
- Components identification
- Feasibility studies
- Software design and development by expert developers/programmers, unless accompanied by detailed written design and analysis of the purpose or context of application
- General-purpose data collection for documentation purposes

From the foregoing, it is clear that simple information gathering, organization of data/facts, pitches and arrangement or modification of existing knowledge would generally not constitute a research study. Thus, an idea or undertaking is not regarded as research if it lacks philosophical and methodological research design.

1.8   Philosophical Research Design

In a later chapter, we give a more extensive presentation on research philosophy and research design. It is important to note that where research lacks a rigorous methodological design, it cannot lead to structured findings that have either qualitative or quantitative attributes. Any processes or behaviour analysed, and any correlations made between particular events, may lack deep and distinctive questioning or investigative thoroughness. Consequently, in the absence of qualitative or quantitative design, there may be no evidence to support data collection and analysis. In quantitative research, the philosophical research design can generate experimental, correlational and survey-based research. For instance, statistics generated for investigation can be used to derive and establish relationships between variables or to determine the extent of correlations (Maiwada & Lawrence, 2015). On the other hand, qualitative research focuses on quality attribute, related to events or phenomenon under investigation, without emphasis on quantitative attributes.

1.9   Some Types of Research Applicable in Information Systems and Cybersecurity Research

The purpose of research is to uncover data and to offer data analysis that leads to theory building, or frameworks for practice, through the appropriate application of scientific methodologies. A few of the more important types of research that students in the fields of information systems, computer science and cybersecurity will use are noted below. While all research will include data, analysis and theory, different types of research place emphasis on different value propositions.

1.9.1   Descriptive Research

Descriptive research is limited to presenting key data about a particular phenomenon, where the data will be a significant contribution to our knowledge, but this type of research offers only limited analysis and a set of simplistic recommendations and does not engage in theory building. The value of descriptive research lies in the data. This is useful, for example, in the case of describing a phenomenon for the first time or one that is not well known.

1.9.2   Exploratory Research

Exploratory research seeks to investigate an unfamiliar problem, where only limited data and limited theory have been published on the subject and where both data and theory are important outputs.

1.9.3   Applied Research

Applied research is conducted to identify the specific characteristics of a real-world problem and to advance solutions which may be action-oriented or policy-oriented.

1.9.4   Experimental Research

Experimental research is designed to assess the effect of one particular variable on a phenomenon, by keeping the other variables constant or controlled.

1.9.5   Theoretical Research

Theoretical research (or basic research) is undertaken for the purpose of examining and exploring ideas, in cases where existing theory is weak or incomplete or requires further work and extension. It is not aimed at offering solutions to a problem in the short term but can contribute to the work of researchers who are problem-solvers. It is highly analytical and may use both deductive and inductive reasoning and counter-inductive reasoning.

1.10   Quality in Research Writing

Once research results have been determined, the next concern is documentation. The quality of academic writing is a critical success factor for CI, IS and CY research. For example, Rathi and Given (2010) discussed various principles that can be applied to scholarly writing in the CS, IS or CY research domain. In an effective written research report, or dissertation, or thesis, a combination of written text, tables and figures, graphs, diagrams and other visualizations can be used. In the 2020s, visualization software enables the researcher to draw highly complex diagrams to showcase the power of the ideas generated by research.

Good research writing will facilitate thesis development by helping the researcher achieve coherence within the context of the study. It is important to note that there is no shortcut to the generation of an acceptable research area, or topic, or research proposal or final research paper. Effective writing must come from research capabilities acquired, such as methodological skills, energy and imagination.

Jalongo (2002) highlights the characteristics of publishable works, which include audience (diverse readership), voice (communication pattern), focus (precise scope/dimension), title (specific title conveying content, purpose and audience), organization (subheadings, visual materials and guidelines), format (template guideline), readability/fluidity, introduction and conclusions.

Below we list ideas that should be used to promote the quality of scholarly writing:

- Understanding emerging trends while identifying a research problem that has wide interest
- Directed use of the Internet as an effective research tool: Conducting an extensive literature search from top scholarly publishers such as Elsevier, IEEE Springer, IGI Global and others and using the available electronic databases
- Providing sufficient theoretical motivation using authoritative sources (e.g. journal articles, industry research papers, others)
- Having a good idea of the frontiers of knowledge of the particular field in which the research is situated
- Attending and presenting papers at seminars, webinars, symposiums and conferences related to your research field
- Engaging in discussion with experts in the field
- Leveraging the supervision process to promote quality improvements

Bibliography

Akintoye, S. B., & Bagula, A. (2017). Optimization of virtual resources allocation in the cloud computing environment. In *IEEE AFRICON* (pp. 873–880). https://doi.org/10.1109/AFRCON.2017.8095597

Diovu, R. C., & Agee, J. T. (2018). Data aggregation in smart grid AMI network for secure transfer of energy user-consumption data. *International Journal of Engineering Research in Africa, 35*, 108–124. https://doi.org/10.4028/www.scientific.net/JERA.35.108

Hefang, F., & Zhaoxia, L. (2010). Research and discussion on introduction to computer science and technology teaching based on methodology. Proceedings of the *2010 2nd International Conference on Education Technology and Computer, China,* V4-293–V4-295. https://doi.org/10.1109/ICETC.2010.5529680

Jalongo, M. R. (2002). *Writing for publication: A practical guide for educators.* Christopher-Gordon.

Jannach, D., Zanker, M., Ge, M., & Gröning, M. (2012). Recommender systems in computer science and information systems – A landscape of research. In C. Huemer & P. Lops (Eds.), *E-commerce and web technologies: Proceedings of the 13th international conference, E-C Web 2012.* Springer. https://www.researchgate.net/publication/268393354_Recommender_Systems_in_Computer_Science_and_Information_Systems_-_A_Landscape_of_Research [pre-print].

Javaid, M. I., & Iqbal, W. (2017). A comprehensive people, process and technology (PPT) appli-
cation model for information systems (IS) risk management in small/medium enterprises
(SME). Proceedings of the *2017 International Conference on Communication Technologies
(ComTech), Pakistan*, 78–90. https://doi.org/10.1109/COMTECH.2017.8065754

Jrad, R. B. N., Ahmed, M. D., & Sundaram, D. (2014). Insider action design research a multi-
methodological information systems research approach. Proceedings of the *2014 IEEE Eighth
International Conference on Research Challenges in Information Science (RCIS), Morocco*,
1–12. https://doi.org/10.1109/RCIS.2014.6861053

Jungmann, R., Baur, N., & Ametowobla, D. (2015). Grasping processes of innovation empirically.
A call for expanding the methodological toolkit. An introduction. *Journal of Historical Social
Research, 40*(3), 7–29. https://doi.org/10.12759/hsr.40.2015.3.7-29

Kolp, M., Snoeck, M., Vanderdonckt, J., & Wautelet, Y. (2019). An overview of scientific areas
in research challenges in information science. Proceedings of the *2019 13th International
Conference on Research Challenges in Information Science (RCIS), Belgium*, 1–5. https://doi.
org/10.1109/RCIS.2019.8877029

Li, Q., Pan, L., & Xiang, Y. (2010). The study on the promotion of work mechanism of students' scientific researches for the school of computer science. Proceedings of the *2010 5th
International Conference on Computer Science & Education, China*, 409–411. https://doi.
org/10.1109/ICCSE.2010.5593597

Liu, K., Jiang, J., Ding, X., & Sun, H. (2017). Design and development of management infor-
mation system for research project process based on front-end and back-end separation.
Proceedings of the *2017 International Conference on Computing Intelligence and Information
System (CIIS), China*, 338–342. https://doi.org/10.1109/CIIS.2017.55

Long, H. (2014). An empirical review of research methodologies and methods in creativity studies
(2003–2012). *Creativity Research Journal, 26*(4), 427–438. https://doi.org/10.1080/1040041
9.2014.961781

Maiwada, S., & Okey, L. E. (2015). The relevance and significance of correlation
in social science research. *International Journal of Sociology and Anthropology
Research, 1*(3), 22–28. https://www.eajournals.org/journals/international-
journal-sociology-anthropology-research-ijsar/vol-1issue-3-november-2015/
the-relevance-and-significance-of-correlation-in-social-science-research/

Molina-Azorin, J. F. (2016). Mixed methods research: An opportunity to improve our studies and
our research skills. *European Journal of Management and Business Economics, 25*(2), 37–38.
https://doi.org/10.1016/j.redeen.2016.05.001

Mulhanga, M. M., Lima, S. R., Massingue, V., & Ferreira, J. N. (2014). Expanding scientific knowledge frontiers: Open repositories in developing countries supported by NRENs. In Á. Rocha,
A. Correia, F. Tan, & K. Stroetmann (Eds.), *New perspectives in information systems and
technologies* (Advances in intelligent systems and computing series, Vol. 275, pp. 127–136).
Springer. https://doi.org/10.1007/978-3-319-05951-8_13

Nnabude, P. C., Nkamnebe, A. D., & Ezenwa, M. O. (2009). *Readings in research methodology
and grant writings* [School of Postgraduate Studies, Nnamdi Azikiwe University (Unizik)].
Rex Charles Patrick Ltd.

Oates, B. J. (2006). *Researching information systems and computing*. SAGE Publications.
https://books.google.co.za/books/about/Researching_Information_Systems_and_Comp.
html?id=ztrj8aph-4sC

Okafor, K. C. (2019). Dynamic reliability modelling of cyber-physical edge computing network. *International Journal of Computers and Applications, 1–10*. https://doi.org/10.108
0/1206212X.2019.1600830

Rathi, D., & Given, L. (2010). Research 2.0: A framework for qualitative and quantitative research
in Web 2.0 environments. Proceedings of the *2010 43rd Hawaii International Conference on
Systems Science, USA*, 1–10. https://doi.org/10.1109/HICSS.2010.317

Raunio, M., Pugh, R., Sheikh, F. A., & Egbetokun, A. (2019). Introduction: Importance of methodological diversity for innovation system studies. *African Journal of Science, Technology,
Innovation and Development, 11*(4), 465–467. https://doi.org/10.1080/20421338.2018.1530406

Roig, M. (2013). *Avoiding plagiarism, self-plagiarism, and other questionable writing practices: A guide to ethical writing.* https://ori.hhs.gov/images/ddblock/plagiarism.pdf

Sacred Heart University Library. (2019). *Organizing academic research papers: Quantitative methods* [online]. https://library.sacredheart.edu/c.php?g=29803&p=185930

Simon, A., Berges, M., & Hubwieser, P. (2016). Different perceptions of computer science. Proceedings of the *2016 International Conference on Learning and Teaching in Computing and Engineering (LaTiCE), India,* 14–18. https://doi.org/10.1109/LaTiCE.2016.1

University of the Witwatersrand. (n.d.). *Plagiarism, citation and referencing styles.* https://libguides.wits.ac.za/plagiarism_citation_and_referencing/Avoidance; https://libguides.wits.ac.za/plagiarism_citation_and_referencing/plagiarism_policy

Van Laar, E., van Deursen, A., van Dijk, J., & de Haan, J. (2017). The relation between 21st-century skills and digital skills: A systematic literature review. *Computers in Human Behavior, 72,* 577–588. https://doi.org/10.1016/j.chb.2017.03.010

Yin, L., Zhang, A., Ye, X., & Xie, X. (2019). Security-aware department matching and doctor searching for online appointment registration system. *IEEE Access, 7,* 41296–41308. https://doi.org/10.1109/ACCESS.2019.2904724

Chapter 2
Computer Science (CS), Information Systems (IS) and Cybersecurity (CY) Research

2.1 Introduction

In this chapter, we dissect and situate the computer science (CS), information systems (IS) and cybersecurity (CY) research disciplines, to assist new researchers to understand how to focus their research efforts. In contemporary research, the line between CS IS and CY research studies is not completely blurred. There are subtle differences, intersections and overlaps. While their research settings may be unique, new researchers need to understand the distinction between CS, IS and CY research enquiries, as well as how they interplay with each other. Traditionally, CS research leans more towards scientific investigations using complex mathematical theories and proofs, to solve problems or to provide answers to questions (Hevner et al., 2004). IS research, on the other hand, tends towards an application of CS to solving real-world problems concerning business, government or society. CY study is concerned with new knowledge in improving and/or developing products relevant to digital security, safety and privacy.

While in many cases, CS will be focused on experimental work and conceptualization of theories, assumptions and constructs, IS seeks to apply those theories and constructs within organizational, economic or social environments (Vahidov, 2012). However, the commonality among the three is that the disciplines may involve design-type research (a series of structured, theoretical and practical activities) (Hevner & Chatterjee, 2010), or design and creation research (Oates, 2006), as well as a product or artefact as a research outcome. Thus, CS, IS and CY can be said to be distinctive fields of study that require common methodological, scientific and technical knowledge sets and in some cases require 'proof of concept' by design and implementation (Ahmed & Sundaram, 2011) to validate the research work.

The intersection among them is that they attempt to create a novel, stimulating knowledge environment, through research enquiry relating to computerization, or digitalization, and their applications to business and society. They use scientific

© The Author(s), under exclusive license to Springer Nature Switzerland AG 2023   13
U. M. Mbanaso et al., *Research Techniques for Computer Science, Information Systems and Cybersecurity*, https://doi.org/10.1007/978-3-031-30031-8_2

philosophy, design principles and critical, creative and innovative thinking to create things that have utility value. In some cases, attempting to draw lines between them can be confusing as there are significant, noticeable overlaps and interconnectivity. In many cases, it is the outcome of academic research that provides the industry with the necessary building blocks to create state-of-the-art commercial products, helping to disrupt the normative landscape of enterprise use of previous innovations.

2.2   CS Research

Traditionally, computer science focuses on theories, assumptions and the engineering of computing artefacts; hence, computer scientists focus on computational theories, mathematical algorithms, data structures, programming structures and algorithms, as well as on concepts that involve computing. The notion is that computer scientists tend to understand the why and how questions that underpin computing models and constructs (Vahidov, 2012). Researchers in this field concentrate on new theories, hypotheses and concepts, using advanced mathematics, algorithms and data structures to invent new approaches that can enhance the internal processes of computing, in an attempt to improve performance and efficiency.

Computer science researchers, therefore, are interested in the broad understanding of the fundamentals of diverse programming language theories, discrete and linear mathematics and operating systems, as well as software design and development methodologies. CS research further entails research enquiry into complexity theory, computational learning theory, algorithm and data structure design, geometric computing, parallel algorithms, cryptography, computational logic and matrix computations. Additionally, considerable research efforts in postgraduate degree programmes at African universities are geared towards computer architectures, compiler optimizations, embedded systems, bioinformatics and computational biology, data mining, database design, graphics and visualization, natural language processing, medical imaging analysis, high-performance computing, human-computer interaction (HCI), networks, distributed systems, quantum computing as well as robotics, machine learning and artificial intelligence and software engineering theory. Figure 2.1 depicts the mapping of the CS research domain. It shows the various branches of research focus and illustrates that theories, assumptions, concepts and mathematical theories can interplay with these branches.

From this diagram, it is evident that research in the field of computer science is very broad. Original research focuses on the creation of fresh formulations of theories, frameworks and approaches to advance the quality of computing artefacts and the broader knowledge base needed to address contemporary problems and digital innovations. The constructs, concepts and theories are advanced to simulate models, implement and test them and to influence computing attributes such as correctness, reliability, readability, computational efficiency and effectiveness (Oates, 2006).

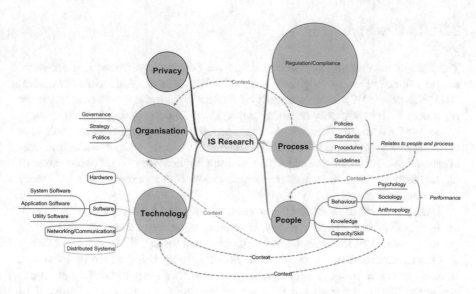

Figure 2.1 *Conceptualization of Computer Science Research*

However, it can be argued that some of the computer science research falls under information systems research, and that, obviously, can be true. The important point is that both these types of research share some conceptual characteristics, but computing is more involved with understanding the *how* and *why* of various computing phenomena, for example architectural structures, processes, operational context and interaction with diverse elements, in the process of performing computing tasks. Therefore, it can be said that the computer science researcher is mainly concerned with computing theories, concepts and assumptions based on mathematical principles, which is the language of computing (Vahidov, 2012). Thus, CS research is primarily concerned with discovering new knowledge in computing phenomena, with strong theoretical foundations and concepts, applied to distinct application domains. In so doing, it concentrates on generating effective theories, models, constructs, algorithms, hardware optimization, data structures, etc., which can advance the computing experience. Consequently, a researcher in CS should gain a more thorough and complex knowledge of the theories and assumptions that underlie the design of computers and computational processes, with an added in-depth understanding of the mathematical and engineering theories and concepts. So, CS research can be said to fall within the ambit of pure design science research (Hevner & Chatterjee, 2010) or design and creation research (Oates, 2006). Designing algorithms to be incorporated into the creation of digital games and other software applications is an example of design and creation research.

2.3   IS Research

The field of information systems interconnects business operations and computer science, focusing on solving real-time business problems (Ahmed & Sundaram, 2011; Vahidov, 2012). Furthermore, it cuts across business and academic/scientific practice, which is primarily concerned with generating computing and technology-based solutions, to solve everyday business problems facing organizations, in a variety of industry sectors (Hasan, 2003). However, IS in its uniqueness emphasizes organizational missions and objectives and the applicability of the wide range of digital technologies to enhancing organizational productivity and efficiency (Thevaranjan & Ragel, 2017).

The main thrust of IS research is to continuously find new ways to improve productivity, efficiency and profitability, leveraging the computational power of computing systems. But IS research must be grounded in theory, either in a novel theory that the researcher is proposing or in an existing theory being advanced to solve a business problem. In some instances, the researcher is simply illustrating the applicability of theory or testing the theory with empirical data. Customarily, if IS research is viewed from the context of a business environment, it encompasses the organization, processes, people and technology (Vahidov, 2012). This perspective is relevant in typical IS research. It implies that the IS researcher should understand the fundamental elements of IS and their relationship within the framework of organizational contexts and the ecosystem of the business. The knowledge of organizational settings, business processes and the people element is vital in guiding the researcher in the choice of a strategy, methodology and philosophy that underpins a typical enquiry. However, the understanding of the relationships that exist among these elements is equally a determining factor for the findings of research work. For instance, Hevner et al. (2004) argue that two distinct paradigms are foundational to IS research – behaviour science and design science, which in effect can be combined in the research process.

IS research is improving how both CS, IS and CY affect society and organizational environments. IS research can be argued to be more versatile, cutting across many more disciplines than computer science. There is consensus that IS research attempts to focus on the effective and efficient utilization of ICT to the benefit of individuals, groups, communities, organizations and society (Lipaj & Davidavičienė, 2013; Munirat et al., 2014). This conception strongly implies that technology, people and process are at the centre of IS research. IS enquiry spans and involves engineering science, management science, social science, natural science, agricultural science, medical science, legal science, political science and other fields and disciplines, depending on the context of the study. Hence, IS enquiry is multidisciplinary, particularly, in the context of behavioural science – human actions in the context of operations and management of IS (Ahmed & Sundaram, 2011), which is the focus of the people and process elements. Conversely, the people element of IS research connects to the behavioural sciences, including the fields of sociology, social and cultural anthropology, psychology and cognitive sciences (Indulska & Recker, 2010; McLaren & Buijs, 2011). What may be lacking is that there is limited behavioural science research in IS research and design, as many researchers concentrate on the technological aspects but fail to attend to human aspects as a design feature. Figure 2.2 depicts the conceptualization of IS research used in this book, illustrating third-order element connections using the mind map[1] approach. Note that a research group can use mind mapping to frame and scope the researchable areas. Finally, the context and environment of the research settings must influence the input and output of the IS enquiry.

Because of the multidisciplinary nature of the field, a typical IS research project may involve hybrid forms of research methodology (Vahidov, 2012). Typically, empirical IS research can span traditional research approaches, including design science, case studies, grounded theory, ethnography and action research, using techniques such as interviews, observations and focus groups (McLaren & Buijs, 2011; Vahidov, 2012). For instance, a usability study of a new artefact can use observations or focus groups to evaluate users' perceptions about the artefact and how easy it is for the users to navigate the different interfaces of the system.

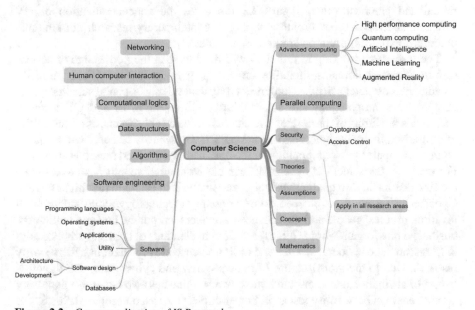

Figure 2.2   *Conceptualization of IS Research*

---

[1]A mind map is a visual tool that conceptualizes and structures our thoughts in an attempt to develop new ideas and graphically represents the ideas, discussed in more detail in Chapter 3.

2.4   Cybersecurity (CY) Research

Organizations are increasingly dependent on cyber-infrastructure, generating increasing cyber vulnerability, making the related field of cybersecurity a major field of study. Cyber-infrastructure is multifaceted, with a broad network of complex objects, all of which are potential points of network vulnerability. Cyber risks and opportunities for bad actors, occasioned by rapid, ever-changing technological advances and human ingenuity, are constantly on the rise. Cyberspace is now so intricate, with the interconnection of small physical objects, artificial intelligence, machine learning and robotics, making security challenges increase exponentially. Consequently, cybersecurity research is taking interdisciplinary and multidisciplinary approaches in attempts to study the technological, security, safety and privacy phenomena. Cybersecurity research is relevant in organizational contexts and relates to both CS, IS and CY research in particular ways. Demand for cybersecurity research is growing; thus, investigations to fill the global cybersecurity void fall well within the ambit of academic and industry research.

CY research is both interdisciplinary and multidisciplinary in approach and will require combined research teams from across disciplines and in specific research areas to work on complex problems. CY research also requires a basic understanding of a few foundational fields of knowledge, to produce research that is of theoretical and practical value. Figure 2.3 illustrates the conceptualization of CY research, showing a typical mind mapping of contemporary research thinking and the inter-relationships among key research themes.

The researcher working in this field will need to span the knowledge of several different research domains, including knowledge previously relevant only to mathematicians, computer scientists, engineers, physical scientists, social scientists, psychologists, anthropologists and so on. Unlike CS research, IS and CY research transcends the limits of disciplines and can foster the establishment of communities of multidisciplinary designers and researchers. Cybersecurity research can be qualitative, quantitative or a combination of the two, namely, a mixed methods approach. For instance, assessment of cyber risks can use a quantitative approach to estimate the risk factors but use qualitative design and analysis to evaluate the threat factors. Cybersecurity should not be seen from the prism of technology alone, rather organization, process, people and governance (management, policy, law and regulation, business ethics, organizational ethics) are all vital elements of cybersecurity. Similar to IS research, behavioural science and design science are prevalent in cybersecurity research. Due to the global nature of cybersecurity and cybercrime, and the challenges of attribution and anonymity, the law discipline has also become a necessary component of cybersecurity research. For example, research on proper attribution of the source material (by one author of another's work or by one designer of another's work) may combine technology and legal perspectives to fully understand attribution issues. Another vital element of cybersecurity research is the focus on risk and resilience, which requires continuous research study to find better ways to ensure the security and safety of cyber infrastructure and other assets. In this context,

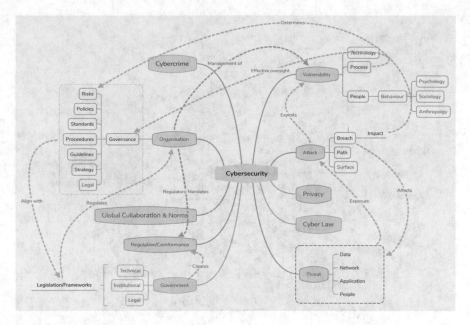

Figure 2.3   *Conceptualization of Research in Cybersecurity*

cybersecurity maturity models must be designed and applied in ways that enable cybersecurity controls to promote resilience and maturity, for proactive security, as opposed to reactive security.

2.5   The Intersection of CS, IS and CY Research

Our discussion so far shows that the domains of computer science, information systems and cybersecurity, although related, are not the same area of study. It can be mystifying to many researchers where the boundaries lie; however, our explanations above attempt to make the differences clear. It is equally important to understand the blurring of the lines, and the intersections, between the three disciplines. While it can be argued that IS research concentrates on the use of products, concepts and theories developed by computer science, the IS researcher must understand some foundational concepts of computer science to focus on the applicability aspect for solving business problems. Conversely, a computer scientist seeking to push the limits of computing knowledge to develop new more relevant theories needs to understand how established theories and concepts attempted to solve technology-related business challenges and what the effects were in practice. The above statement attempts to correct a common misunderstanding that computer science research relates to the professional use of computing technology. Rather, it is IS research that relates to everyday computer use.

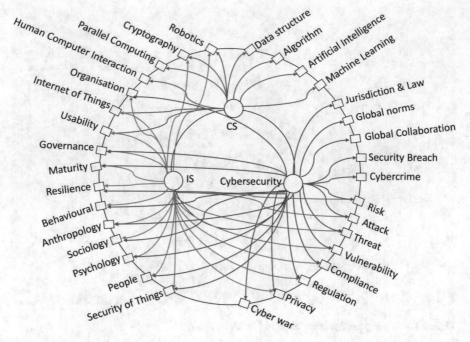

Figure 2.4   *Mapping of CS, IS and CY Relationships*

To put it simply, while information systems research concentrates on the use of computers, and the organization in which IS is used, computer science is largely concerned with theories, concepts and principles that underpin everyday computing and information technology. On the other hand, cybersecurity spans computer science, information technology and other sciences (Vahidov, 2012). Figure 2.4 depicts the mapping of relationships of the three disciplines to each other, illustrating some of the intersections and variances.

Multidisciplinary research examines a phenomenon from the perspective of two or more disciplines (but not so many that it makes the work cluttered or impossible to pursue). Interdisciplinary research searches for the niches, new insights and new fields of knowledge exploration at the points where disciplines overlap, rather than trying to stretch insight and analysis across disciplines.

Thus, in the field of cybersecurity, multidisciplinary research could involve topics that include a technological (software engineering) and a regulatory perspective (law and engineering), or a socio-technical perspective, such as questions of algorithmic bias. Interdisciplinary research would focus on the design of an integrated cybersecurity framework to use computational mathematics and other techniques in computer science to predict and respond to cybersecurity breaches.

Hence, IS and CY research are more crosscutting, interdisciplinary and multidisciplinary than computer science, with cybersecurity at the top of the curve, as illustrated in Figure 2.5.

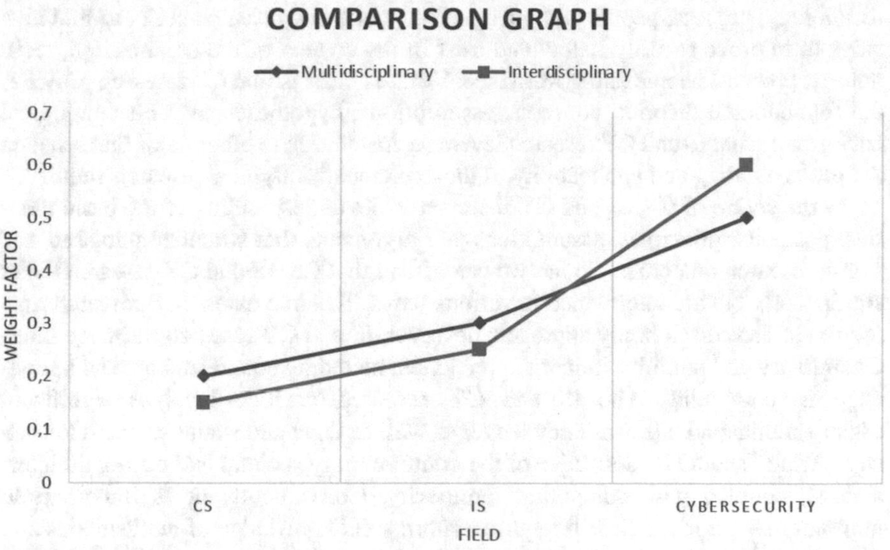

COMPARISON GRAPH

Figure 2.5   *Comparison Chart Across Three Fields of Study*

The y-axis represents weighting estimates, based on the characterization of multidisciplinary and interdisciplinary factors in the context of CS, IS and CY research, noting that this is just a way of enabling our thinking. Insight suggests that the practice of CY research cuts across many more disciplines than CS research, as depicted in Figure 2.4, though this may change as computational modelling becomes more powerful in the data science domain.

In practice, there exist even more intersections of CS, IS and CY research than are presented in Figure 2.4. A good example of intersection is an Internet of Things (IoT) usability study, which can span CS, IS and CY. However, research on IoT can also be addressed from only one of the perspectives of CS, IS or CY, depending on the issues or problems. Preferably, researchers in the IS and CY fields will need to study and understand the fundamentals of CS to situate particular research problems within the context of particular conceptual and theoretical analytical frameworks. In addition, a researcher in cybersecurity needs to gain extensive background knowledge of applicable theories, concepts, frameworks, assumptions and arguments, in order to progress the research enquiry productively. Cybersecurity researchers must also become familiar with their particular sector of application, whether banking and finance or public health or e-commerce.

2.6   Summary

So, we can say that CS research concentrates on finding new computing theories, concepts and paradigms to continuously advance computing experience and efficiency, while IS research puts these new ideas in the context of the organization, technology, process, people and governance, while CY research seeks to find new ways to improve security, safety and trust in the context of the organization, technology, process, people and governance. What is clear is that CS research provides the foundational theories, concepts, assumptions, hypotheses and epistemological background that IS and CY research leverage, in addition to other fields that broaden the understanding and applicability of those elements in typical research enquiry.

At the centre of CS, IS and CY research is the understanding of the basic theories, concepts, principles, assumptions and arguments that stimulate unbiased and sound research outcomes. So far, we can affirm that CS, IS and CY research differ significantly but are intertwined in various ways. To some extent, CS research may require an interdisciplinary approach; however, IS and CY research must use interdisciplinary and multidisciplinary approaches, including sound background knowledge in computing. The IS and CY research practice involves significant organizational and business knowledge, as well as legal and social sciences knowledge, which should be derivative of the contemporary societal and economic interactions, coupled with computing familiarity. Consequently, it is important to pinpoint that CS research will require a more solid knowledge of mathematics and theoretical formulations than IS and CY research. We can see CS, IS and CY research as a tripartite track that is distinct in some ways but closely related in other ways. This leads to our main conclusion for this chapter, namely, that there is a need for a significant coupling of CS, IS and CY research studies in the academic and industry environments.

Bibliography

Ahmed, M. D., & Sundaram, D. (2011). Design science research methodology: An artefact-centric creation and evaluation approach. *ACIS 2011 Proceedings - 22nd Australasian Conference on Information Systems, Australia.* https://aisel.aisnet.org/acis2011/79/

Hasan, H. (2003). Information systems development as a research method. *Australasian Journal of Information Systems, 11*(1), 4–13. https://ro.uow.edu.au/buspapers/458/

Hevner, A., & Chattarjee, S. (2010). *Design research in information systems: Theory and practice.* Springer. https://www.springer.com/gp/book/9781441956521

Hevner, A., March, S., Park, J., & Ram, S. (2004). Design science in information systems research. *MIS Quarterly, 28*(1), 75–105. https://www.researchgate.net/publication/201168946_Design_Science_in_Information_Systems_Research

Indulska, M., & Recker, J. (2010). Design science in IS research: A literature analysis. In S. Gregor & D. Hart (Eds.), *Information systems foundations: The role of design science.* ANU E Press. https://doi.org/10.22459/ISF.12.2010.13

Lipaj, D., & Davidavičienė, V. (2013). Influence of information systems on business performance. *Mokslas – Lietuvos Ateitis/Science – Future of Lithuania, 5*(1), 38–45. https://doi.org/10.3846/mla.2013.06

McLaren, T., & Buijs, P. (2011). A design science approach for developing information systems research instruments. Proceedings of the *6th International Conference on Design Science Research in Information Systems and Technology, USA,* 1–10. http://www.rug.nl/staff/p.buijs/design_science_approach_for_developing_isr_instruments.pdf

Munirat, Y., Sanni, I. M., & Kazeem, A. O. (2014). The impact of management information system (MIS) on the performance of business organisation in Nigeria. *International Journal of Humanities Social Sciences and Education, 1*(2), 76–86. https://www.arcjournals.org/pdfs/ijhsse/v1-i2/8.pdf

Oates, B. J. (2006). *Researching information systems and computing.* SAGE Publications. https://books.google.co.za/books/about/Researching_Information_Systems_and_Comp.html?id=ztrj8aph-4sC

Thevaranjan, D., & Ragel, V. (2017). The impact of management information system on service quality. *International Journal of Research, 4*(8). https://journals.pen2print.org/index.php/ijr/article/view/8358/8109

Vahidov, R. (2012). Research in information systems. In R. Vahidov (Ed.), *Design-type research in information systems* (pp. 51–75). https://doi.org/10.4018/978-1-4666-0131-4.ch003

Chapter 3
Designing the Research Proposal or Interim Report

3.1 Introduction

The research proposal is an important milestone in the overall process of postgraduate research. Various schools may call this document by different names, noting that it may also be called the interim report in your institution. Some institutions have a short and a long proposal. Whichever approach your institution uses, the purpose of these documents is similar. We will deal mainly with the research proposal, and we will also refer to the interim report, which is a little more developed than the proposal.

The research proposal is presented for assessment for the following degree programmes, for the Masters by coursework and research, for the Masters by dissertation only and for the PhD programme. Its role is to provide the basis on which to ascertain the student's understanding of research design and assess whether the research can be completed within the constraints of time and resources. The research proposal will be read by an internal or external reader who is not the supervisor or will be reviewed by a proposal panel. The reader(s), or the panel, will provide a report and guidance as to whether any minor or substantial changes should be made to the research design. The research proposal is generally the foundation for the first three chapters of the study, depending on the specific requirements for the degree programme. It includes a short introduction, as well as the background, literature review and methodology sections. A poorly designed proposal will not pass this milestone and may result in deregistration from the degree programme. This chapter discusses the components of the research proposal design, needed to successfully prepare and present a proposal of acceptable quality. The long proposal or interim report will require more advanced development of the relevant sections; see Figure 3.1.

Some schools require a research concept paper, noting that this is an initial short paper, of no more than five pages, which is the basis for the application to the degree programme. This is simply a description of the topic proposed by the applicant, to enable potential supervisor(s) to evaluate the applicant's idea, scope and

U. M. Mbanaso et al., *Research Techniques for Computer Science, Information Systems and Cybersecurity*, https://doi.org/10.1007/978-3-031-30031-8_3

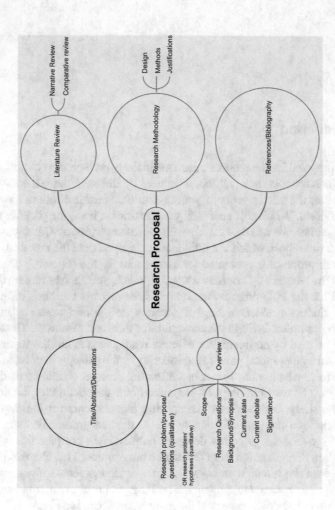

Figure 3.1 *Content of the Research Proposal. Note.* Above is a typical outline that presents a general idea of the content items in the proposal

significance of the topic. The applicant should demonstrate that the topic is research-able and is achievable within a specific timeframe. The research concept paper briefly describes the general area of study, sharing the current state of knowledge and any topical arguments on the subject. It should provide a rough idea of the fea-sibility, approach and methodology, timeline and the originality of the proposed research. We do not cover the research concept paper in this chapter.

3.2   Visualizing the Research Proposal as Motivation to Conduct Research

A research proposal is a coherent document presenting (i) a very short introduction, (ii) the research problem statement and purpose statement and (iii) the research ques-tions. It includes (iv) a background discussion, (v) a literature review and analytical framework and (vi) a methodological discussion (Figure 3.2). The interim report is a longer, more developed version of the research proposal, longer in word length, with more substance in the research design, offering a more developed literature review and methodology section. The interim report typically includes the first three full chapters, whereas the research proposal does not have any completed chapters.

Within the context of the research process, a good proposal must capture the necessary information to enable the readers to quickly complete the assessment. In the computer science (CS)/information systems (IS) and cybersecurity (CY) domains, rapid advances in research design are occurring on a daily basis. Therefore, a good proposal needs to clearly state what is to be achieved, the motivation for the work, as well as a clearly mapped-out strategy to achieve the research objective. In particular, the departmental Postgraduate Committee will expect PhD researchers to present good research proposals as the basis for their candidature. The proposal must be successfully defended, and ethics clearance requirements must be met, before the experimental work, or data collection can take place. Figure 3.3 shows a visualization of the proposal relevant to the CS, IS and CY research areas.

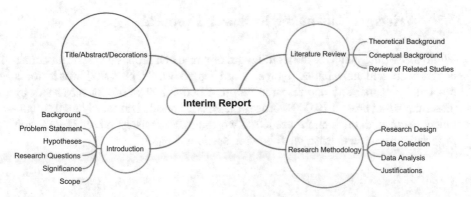

Figure 3.2 *Content of the Interim Report*

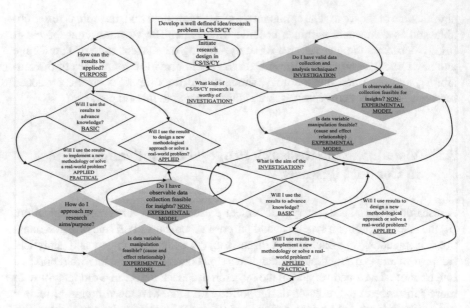

Figure 3.3 *Research Proposal Visualization*

From the visualization in Figure 3.3, the following are pertinent:

- Designing the research problem and purpose which can be basic, applied or experimental in nature
- Presenting the existing knowledge spectrum on the subject matter and selected key insights into the evolution of the ideas (background and literature review)
- Presenting a research methodology suitable to the CS/IS/CY and related knowledge domains
- Demonstrating research feasibility, originality and significance of the research idea

3.3 Writing a Convincing Research Proposal

While the requirements for the word length of a research proposal will differ across departments and universities, typical word length would be 4,000 words for a Masters by coursework and research report, 6,000–8,000 words for a Masters by dissertation and around 10,000 words for a PhD proposal. This word length is sufficient to cover all the sections needed, if words are used wisely. A concise research proposal must have, at least, the following sections:

3.3.1 Cover Pages

The cover pages for the research proposal should include a title page, the plagiarism declaration, a list of abbreviations, a list of acronyms, a glossary, then the table of contents, a list of figures, a list of tables and a list of annexures. Each institution will have its own guidelines, but the requirements are very similar. The simplest approach is to follow the APA style guide for the formatting of research papers, noting the popular style guidelines: paper format, available at https://apastyle.apa.org/style-grammar-guidelines/paper-format. The webinar titled A Step-by-Step Guide for APA Style Student Papers, available at https://apastyle.apa.org/instructional-aids/tutorials-webinars, provides useful information about presenting a well-structured, well-organized research paper.

3.3.2 Research Introduction

The introduction should be no more than one or two paragraphs, stating what the research field is, what the research topic is and what the specific nature of the enquiry is. It should not explain this in detail that will come in the sections that follow. The introduction is distinct from the longer background section. Some attributes of a good introduction include the following:

- Provides the cynosure (i.e. placing the research idea at the centre of the reader's attention).
- Provides a very brief but thought-provoking idea (hook) of why the research is interesting.
- States whether the research is quantitative or qualitative in nature.
- Being persuasive, as well as inspiring a supervisor, or an evaluator, to read the proposal, is the key to successful proposal submission.

3.3.3 Research Problem Statement: Identify a Research Problem That Needs Attention

Regardless of the nature of the localized or broader problem, the attention of a supervisor or a reader is drawn once a specific need, or gap in knowledge, is clearly identified. By demonstrating a high level of understanding of the research problem, the researcher must trigger agreement on proceeding to complete the research. For the research problem statement, the key elements include (a) clear problem description in relation to research topic context and (b) gap in knowledge briefly stating why the research is necessary. The problem description should include the main problem and at least three (Master's level) elements but not more than five (PhD level) elements of the problem.

3.3.4 Research Purpose Statement: Clearly Stating the Purpose of the Study

The key components include (a) a well-structured and clear statement explaining the purpose or objectives of the study. The main purpose is usually formatted as 'The purpose of this research is to investigate …'. As in the case of the problem statement, this should be followed by clearly highlighted sub-objectives, mirroring the elements of the research problem. The elements of the research problem are expressed as sub-objectives in the purpose statement. In both qualitative and quantitative research designs, it is important that there is consistency between the elements of the research problem and the sub-objectives of the purpose statement. This gives the research design a solid foundation.

3.3.5 Research Questions and/or Hypotheses

In qualitative research, you will present a main research question that frames the overarching research problem, as well as three (or more, but not more than five) research sub-questions that frame the elements of the research problem. If there are three elements of the research problem, there should be three research sub-questions. If there are five elements of the research problem, there should be five research sub-questions. In quantitative research, the hypothesis formulation can be fairly extensive, setting out three or more hypotheses, each with a few hypothetical statements. In mixed methods research, the design may include a few hypotheses to be tested and a few research questions to be examined through more in-depth research methods.

3.3.6 Brief Statement of Methodology

This research design section can include a very short statement of methodology, noting that an extensive discussion of methodology comes later, after the major background discussion and literature review. This brief methodology statement can simply state the methodological approach to be used, for example, qualitative analysis, software design and data modelling.

3.3.7 Statement of Significance

Over and above the problem statement, a research proposal must offer clear justification for embarking on the research, by presenting valid reasons for tackling the problem. Building the justification could include the following:

- Providing a few statistical references from reputable sources that reflect the significance of the problem
- A brief explanation of the consequences that would arise from allowing the problem to persist without addressing it
- Highlighting the innovation and expected benefits of the research outcome to potential beneficiaries and how it will impact their institutional and/or economic and/or societal future

For instance, the discussion of significance can highlight how the research could improve productivity, efficiency, profitability and other successful outcomes. Or, the justification can relate to better understanding of specific aspects of the process of digital transformation in a particular context, for example, cybersecurity resilience in the banking sector or systems integration in the emerging fintech sector.

3.3.8 Background to the Research Problem Statement: Including Powerful Supporting Information

It is necessary to include an informative historical overview, technical references, relevant datasets and other supporting details that clearly highlight the background to the research problem statement. The supporting information can include diagrams, algorithms, reference to existing policy documents or law, or regulation, or digital transformation in a particular sector, or other relevant existing knowledge, all with proper in-text references.

3.3.9 Literature Review

The literature review section should provide an overview of the relevant theoretical and practice-oriented knowledge pertinent to the nature of the research problem and its elements. We set out our understanding of existing knowledge in the literature review, in order to demonstrate (a) that we are familiar with the current state of knowledge and (b) that we are conscious of the gap in knowledge. In each new study that we conduct, we build on existing knowledge.

In the research proposal, the literature review is relatively short, possibly 1,500–2,000 words at Master's level and around 3,500 words at PhD level. For the interim report, however, a full literature review chapter of around 6,000 words at Master's level and around 8,000–10,000 words at PhD level is required.

3.3.10   Research Methodology and Methods/Procedures

The research methodology section in a well-written research proposal will include an indication of (a) the broad methodology, whether it is a qualitative or constructivist methodology or whether it is a quantitative, survey-style methodology or a quantitative study with other controlled parameters. The research methodology section should include a clear description of the steps for data collection, the steps for data analysis and an indication of research limitations or unavoidable constraints. We will discuss these components in greater detail in Chapter 6. In the research proposal, the methodology section is relatively short, possibly 1,500–2,000 words at Master's level and around 3,500 words at PhD level. For the interim report, however, a full methodology chapter of around 4,000 words at Master's level and around 8,000 words at PhD level is required.

3.3.11   List of References and In-Text Referencing

The References section must detail the sources of all materials used in the proposal either in MLA (Modern Language Association of America) format, APA (American Psychological Association) format, IEEE (Institute of Electrical and Electronics Engineers) or any other format recognized by the postgraduate board of the researcher's institution. In addition, the entire proposal must cite all the references consulted in the text. A typical set of references for a Masters by coursework research proposal can be around 30 references, for a Masters by dissertation research proposal can be 50 or more references and for a PhD research proposal can be 80 or more references.

3.4   Framework for Thinking About Originality in the Research Design

Originality in the Masters and in the PhD refers to originality in the data, originality in the analysis, originality in the methodology and/or originality in the conclusions and theories generated in the research report, dissertation or thesis. An originality cognitive framework is shown in Figure 3.4.

Originality in research emerges by design, not by accident. The framework above sets out a three-tier approach to achieve originality. As shown in the diagram,

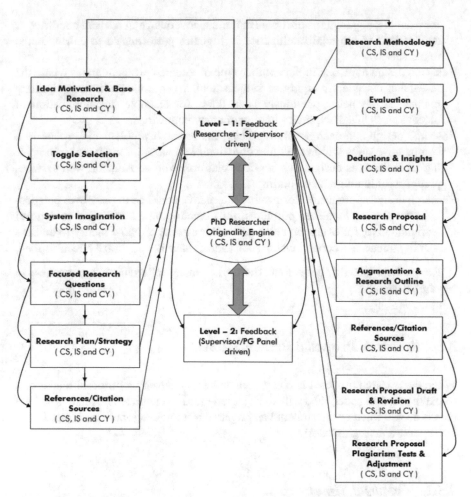

Figure 3.4 *Originality Cognitive Framework*

original research requires an organic thought process, devoid of copied/plagiarized content. The challenge faced by all Masters by dissertation and PhD researchers is to entrench originality in their works. Some suggested approaches that could be used to generate new knowledge include detailed observations, experiments, new techniques to problem-solving and using the systematic literature review technique, discussed in a later chapter. Considering originality, some recommended approaches for effective research design include the following:

Research design approach 1: Observe and identify research problems, for instance, the inconsistencies or gaps in existing studies and analysis, for example, inconsistencies in the discussion of security vulnerabilities in web platforms.

Research design approach 2: Another type of research problem arises where there is an apparent oversight or neglect of an important focus area in digital adoption, for example, applications of robots in human governance in African countries or applications of artificial intelligence in logistics performance to enhance customer service.

Research design approach 3: Yet another type of research problem arises where the researcher considers merging of subject domains in a unique interdisciplinary manner as has never previously been done, for example, the application of mechatronics in a disciplinary field with many boundaries.

Research design approach 4: Alternatively, the research problem may relate to a comparative study within the same domain of knowledge, for example, comparing an indigenous design web services platform with an Amazon web services platform for intelligent computing.

Research design approach 5: Yet another alternative could be to use novel perspectives for solving localized problems, thus creating new knowledge or a novel learning curve, for example, using particular aspects of game theory to address security threats in a context where such aspects have not previously been applied.

Researchers should note that there are many alternative approaches to research design.

3.5   Research Proposal Presentation

When the research proposal is ready, it must be reviewed by a proposal reader or a research proposal panel, who will all be experienced academics. The readers and the panels decide whether the student has prepared a suitable research problem, literature review and research design.

3.5.1   Proposal Timeframes

In most cases, postgraduate students are required to give a presentation of their research proposals within the first year of their candidature. The presentation is required within the first 6–12 months for full-time doctoral students and 12–18 months for part-time students. Masters students may be required to complete their review presentation within the first 6 months for full-time students and first 12 months for part-time students. The student's continuation of candidature is dependent upon the satisfactory confirmation of the proposal review. Though the above scenario varies from one institution to another, the research students usually cannot proceed to data collection until after their research proposals have been reviewed and confirmed by both the supervisor and the review panel.

3.5.2   Research Proposal Presentation and Feedback

Institutions have varying requirements for research proposals in the fields of CS, IS and CY, but the requirements broadly follow those set out in Figures 3.1, 3.2 and 3.3 above. The student should consult with the assigned supervisor to understand where there may be some differences from the approach set out in these diagrams.

If the proposal is sent to a reader, the student will usually be required to present a 1-hour seminar to the supervisor, other academics and students engaged in the study programme. If, on the other hand, the proposal must be presented to a review panel, then the student must present a 30-min presentation, followed by questions and comments. The goal is to provide the researcher with an opportunity for academic conversation.

The proposal reader or the review panel provides a set of formal comments and queries, which must be addressed by the student, who must make corrections and revisions to the proposal.

Many universities currently host research proposal presentations using online tools like Cisco Webex, Google Meet, Microsoft Teams and Zoom, among others. Find out the recommended platform and master the platform prior to the seminar or proposal review panel. Slide presentations of 15–20 slides will usually be required. The presentation should include:

- The title page
- Short version of the research problem statement
- Research questions or hypotheses
- Brief literature review of relevant concepts and theories used to generate the analytical framework
- Methodological design including data collection and data analysis instruments and methods, as well as expected contribution to knowledge

There are limited time slots for proposal presentations, so stick to the time schedule you are given for your presentation. It is good practice to have your presentation and the full proposal document open on your laptop during the presentation. It is important to make notes about the clarificatory questions you are asked and the recommendations made to improve the presentation. Make the relevant revisions immediately after the proposal presentation. You will make minor or significant revisions based on the inputs from the presentation seminar, or proposal review panel, or the proposal reader. All these inputs help to enrich the quality of the research design.

3.5.3   Research Proposal Review Panel Composition

Depending on the institutional requirements, there may be an established proposal presentation and review panel, comprising at least four members. This panel consists of the main supervisor and any co-supervisors, as well as other selected members of academic staff from the department or faculty who have relevant expertise. In addition, an external academic may be invited to serve on the panel. One panel member is usually nominated as the chairperson, who may be the most senior member of the committee. In some cases, postgraduate students are allowed to observe the process and may ask questions but may not participate in the discussions of the academic review panel.

3.5.4   Research Proposal Panel Decision

Research proposals are subjected to rigorous review. Based on the panel discussion of the academic quality of the written proposal and the oral presentation, the chair of the panel will prepare a written report for submission to the faculty administration. The review report will outline strengths and weaknesses, noting specific areas where the proposal is satisfactory or unsatisfactory. Recommendations are made for improvement. Results of the review assessment will be communicated verbally to the student after the panel discussion, at which point the student will be made aware of the recommendations. A written recommendation which denotes a positive outcome will include one of the following options:

- *Full acceptance* of the proposal
- *Conditional acceptance* of the proposal subject to *minor revisions* to the satisfaction of the panel or supervisors
- Recommendation for *major revisions*, leading to a further review of the proposal in 3–4 months, as deemed fit by the review committee

If the proposal is considered *unsatisfactory*, the student will be given the opportunity to revise and present as recommended by panel, probably over a longer period than 4 months. In the case that the second review is also deemed unsatisfactory, the panel may make a recommendation regarding the student's continuation of candidature. Recommendations may include one or a combination of the following: change of supervisor(s), change of topic, downgrading of degree, probation or termination of candidature depending on the prevailing rules guiding the department.

3.6   Closing Thoughts

It is important that the researcher is well prepared for the proposal submission and presentation. In this regard, the following seven guidelines are useful:

Guideline 1: Familiarize yourself with the expectations of the review panel/proposal readers. Discuss this with the supervisor(s) to be sure of the exact requirements. Usually, each of the proposal review panel members is provided with a copy of the research proposal a few weeks prior to the presentation. At some point before defending the research proposal, it is advisable to sit with the supervisor and available domain experts for a preparation session, to organize and plan the presentation. The supervisors have read many proposals and know what committees look for. Having the research proposal presentation structured well in advance will ensure a confident presentation and a valuable learning experience.

Guideline 2: Talk to students who have already presented their research proposal, as they can provide insight into the proposal review experience.

Guideline 3: Attending academic conferences and other proposal presentations will enable you to observe interactions between students and committee members, hear the types of questions that arise and identify the characteristics of strong and weak proposal presentations.

Guideline 4: Practice making the presentation ahead of time, and make adjustments as needed. Getting your friends/colleagues to sit through a practice and make comments can help identify the portions of the presentation that need to be adjusted.

Guideline 5: The proposal must reflect realistic goals.

Guideline 6: For all types of proposals, a well-written proposal must follow a logical sequence of discussion that is coherent, consistent and devoid of errors. Proper attention must be given to content, research design, layout, grammar, spelling, punctuation, diagrams and referencing, all essential to pass this stage. Proofreading and editing your proposal prior to submission are indispensable.

Guideline 7: Plagiarism is not allowed and is a ground for disciplinary enquiry. A student does not need to use anti-plagiarism software, as the student knows whether they have plagiarized. Only the supervisor should submit the proposal to generate the similarity report.

Bibliography

Daoud, S., Alrabaiah, H., & Zaitoun, E. (2019). Technology for promoting academic integrity: The impact of using Turnitin on reducing plagiarism. Proceedings of the *2019 International Arab Conference on Information Technology (ACIT), United Arab Emirates*, 178–181. https://doi.org/10.1109/ACIT47987.2019.8991046

Hao, J., & Ching-Chiuan, Y. (2009). PhD in design: A reflection from a PhD student and his supervisor. Proceedings of the *IEEE 10th International Conference on Computer-Aided Industrial Design & Conceptual Design, China*, 146–150. https://doi.org/10.1109/CAIDCD.2009.5375111

Vrbanec, T., & Meštrović, A. (2017). The struggle with academic plagiarism: Approaches based on semantic similarity. Proceedings of the *40th International Convention on Information and Communication Technology, Electronics and Microelectronics (MIPRO), Croatia*, 870–875. https://doi.org/10.23919/MIPRO.2017.7973544

Chapter 4
Adopting a Funnel Strategy and Using Mind Mapping to Visualize the Research Design

4.1 Introduction

Academic research is more than a project or a piece of writing. A research paper, dissertation or thesis can only be written up after relevant research has been conducted. Research students will have different motivations for conducting a research inquiry, some will have a specific problem they would like to address through experimentation, while others will wish to add to theory. Designing a research study is often a major challenge for novice researchers due to a limited understanding of the research process and possibly also the particular meaning of research (Carlsson, 2006) in the context of a specific research problem. Many academic researchers are uncertain how to identify and craft the research problem to be addressed in the inquiry and how to frame the main research question. The beginner must conceptualize a research idea, set the scope of the research problem and consider whether the problem is researchable (Pajares, 2007). However, finding a researchable area may not be straightforward. This is where the funnel strategy is useful. The funnel strategy is a mental model, which a researcher can adapt, to provide guidance on the research process, enabling the flow of thoughts and ideas throughout the research journey.

4.2 The Concept of Research Strategy

A research strategy is the design of essential steps and activities that directs the view of the study and the resolve to conduct a methodologically sound research inquiry, in order to arrive at a valuable research outcome. It enables the researcher to remain motivated and focused, which can help lessen frustration, and improve the quality

U. M. Mbanaso et al., *Research Techniques for Computer Science, Information Systems and Cybersecurity*, https://doi.org/10.1007/978-3-031-30031-8_4

of the inquiry. Time and resources can be saved by the adoption of an effective research strategy (Hauberg, 2011).

In the section that follows, we introduce the concept of a funnel approach to research strategy that enables the researcher to visualize the study process from the beginning to the end.

4.3   The Funnel Strategy Approach

The most important part of research design is formulating a researchable problem statement that clearly describes and establishes the scope of the study (Hassani, 2017). To do this, the researcher needs to create a logical process flow that commences with the identification of the broad subject area (e.g. cybersecurity (CY)), identifying the overarching research problem (e.g. cyber vulnerability) and three or four elements of the problem (e.g. cyber vulnerability challenges, practices and future needs). The design of the research study must then follow a rigorous processes of identifying the purpose or goals of the study, framing appropriate questions to interrogate the problem and devising ways to measure (using certain metrics for quantitative studies) or understand (using qualitative dimensions for qualitative studies) the phenomenon of study (Salhin et al., 2016). The research design includes choosing the data collection and data analysis method(s); planning, designing and conducting the investigations (through surveys, or interviews, or document review or a combination of these); and ensuring that the methodology enables the researcher to draw valid conclusions (Indulska & Recker, 2010). Furthermore, for the research outcome to be useful, or add new ideas to the body of knowledge, the research enquiry should focus on addressing everyday research problem(s) relevant to the economic and social challenges faced in a particular context or sector.

Ellis and Levy (2009) describe the concept of the funnel in a research study from the perspective of 'research-worthy problem (P)' described as the input to choosing a research area; 'the valid peer-reviewed literature (L)' as 'key funnel (components) that limits the range of applicable research approaches, based on the body of knowledge'; and 'the data (D)' that is relevant to the enquiry, which aids in sifting through the huge available data to find the exact study area.

The funnel strategy described in Figure 4.1 approached the concept from a different outlook, but with a similar thought. This funnel framework describes an initial approach to a research study and its design. Metaphorically, a funnel can be used to guide the thinking of a researcher on where to begin the research process, starting from a wider view of the subject matter under investigation. If we consider the wide mouth of the funnel as the broad field of study, for example, cybersecurity, and the narrow stem of the funnel as the specific research problem, then the research questions and research strategy can be derived from that, to guide the research.

Thus, the funnel framework can help to direct the researcher's thought process about how a research study can be carried out meaningfully in the real world. This is the psychological context of research, where the researcher engages in creating an

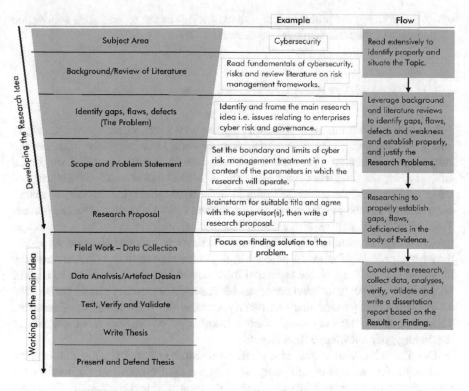

Figure 4.1 *The Funnel Strategy*

understanding of the relationships between the various research phases to direct her/ his intuitive perception. The researcher needs to develop a sound, rational decision-making process, which is logical and evidence-based (MacIntyre et al., 2014). Clearly, a decision-making process can be described as a cognitive progression, capable of directing a course of action in a reasoned sequence, of which the outcome can either be convincing or illogical, based on particular assumptions or presumptions. Simply put, the funnel strategy builds the reasoning skills that are required in a typical research environment, including sense-making, constructivism and other thinking skills.

Sense-making is a way of creating situational awareness and thoughtfulness to appreciate situations of high intricacy or uncertainty and still make informed research decisions. Constructivism means that the researcher actively builds knowledge by constructing a view of the phenomenon based on data gathered from real people, who each have their particular interpretation of the phenomenon being studied (Ernest, 1996). Within the mass of data collected, the researcher grapples to build constructs, combining perspectives on the internal and external dynamics affecting the research problem. The funnel strategy can support this sense-making process, encouraging the researcher to filter the ideas encountered in the literature while constructing the problem under investigation. In the sections that follow, the components of the framework are explained.

4.3.1 The Subject Area

A first step in the framework is identifying the broad subject matter of interest, in which the researcher may have some basic or little knowledge, for instance, the subject of cybersecurity vulnerability. Subsequently, having identified the broad subject field, the researcher must sieve through available scholarly literature, and other publications relevant to the topic, in order to ascertain that the subject matter is an area of interest worth devoting rigorous and time-consuming effort to pursue. This initial effort will require establishing a clear understanding with respect to intellectually stimulating knowledge gaps, or flaws in existing ideas or solutions, through a process of detailed note-taking. The researcher has to understand the ideas trending on the selected topic and must devise how the ideas relate to a set of broader or narrower associated concepts. More so, the researcher should create a list of keywords and phrases that can clearly express the topic under consideration. These keywords can help in advancing knowledge on the topic, which aids in expanding your knowledge foundations.

Building a knowledge tree: This phase of research is a critical stage that lays the foundation for the rest of the study. It serves a dual purpose of (i) helping the researcher to firmly grasp a broader view of the topic area and specifically (ii) ensuring that the focus/scope of the study has not been covered by other researchers or is not part of an ongoing research effort. This initial phase can uncover if there is sufficient information available to aid the investigation, very important for postgraduate researchers. The researcher or student should use this time to build focus, in order to avoid changing the topic. Each topic has its particular challenges, and proper scoping of the work will ensure that consistent progress is made.

4.3.2 Preparing and Writing the Background Discussion

The next step for the researcher is to gain considerable theoretical background and establish the underlying theories and concepts. The researcher should view the theoretical background as the catalyst for understanding, analysing and connecting the factors or variables that influence the study. Background reading enables the researcher to situate the enquiry, to understand the underlying theoretical underpinnings, essentially to ground the theories and concepts that can support the investigation of the phenomenon of study. Theories (e.g. socio-technical theory and theories of regulation) and concepts (e.g. digital innovation and cybersecurity regulation) are vital to explain the influencing factors and establish the domain of your particular research problem, relative to existing knowledge. Furthermore, the researcher must engage with the real-world background data relevant to the problem, if the study includes a practice orientation, for example, analysing the development of broadband infrastructure and services in a selected African country or analysing the digital skills needed for effective cybersecurity operations. Without extensive background reading, the researcher will struggle to explain why the problem exists or to provide motivation for the appropriateness of the enquiry.

Consequently, the background section attempts to offer an understanding of theories and concepts that have relevance to the study and prepares the researcher to scope the problem space and domain. So the goal of the theoretical background is to reinforce the research in a manner that assumptions made can be critically evaluated to ascertain their appropriateness for the study. It serves as a link to existing knowledge, which fosters a better understanding of the research problem (qualitative research) or hypothesis (quantitative research) and propels the choice of research methods (Hassani, 2017). Besides, the theoretical background underscores the generalization of the various elements of the study and phenomenon under investigation. Theoretical backgrounds complement the review of literature in the area, noting that the guiding theories and concepts should be 'planted' in the background discussion and then discussed extensively in the literature review, where the researcher will identify the constraints and limitations of the existing theories and concepts. More so, focusing on specific variables or factors, this initial theoretical background aids the researcher to identify and locate the primary and secondary sources of data relevant to the research. Given the extensive information available online, keywords drawn from the initial reading material provided for lectures can be used to identify relevant sources of background knowledge.

Based on the funnel strategy, narrowing the focus to a specific topic of interest, such as *cybersecurity challenges experienced in a particular economic sector* (banking or medical information systems (IS)) in a particular country (e.g. Ghana, Kenya, Nigeria), the researcher should establish sufficient knowledge of the applicable theories and concepts in computer security, information security, network security, application security, people security, etc. The researcher then narrows the scope further to the factors and variables relevant to *cybersecurity challenges*, in the context of the topic. Subsequently, the researcher may highlight *cyber risk governance, cybersecurity professional skills* and *cybersecurity applications* in the banking sector as the three elements or variables to be examined in relation to cybersecurity challenges. These even more specific concepts can help identify the appropriate literature. In practice, the researcher begins to dig deeper into the particular topic, writing up detailed notes, in this case on *cyber risk governance* in the banking sector. From these notes, the researcher should be able to write a 1000-word background section highlighting the main research problem. Another example of the design of a research problem would be *identity theft in the banking sector*, stating three elements of this problem, for example, (i) types of threats due to possible abuse of personal information, (ii) financial losses and building customer trust and (iii) effectiveness of legislation and regulation in addressing threats arising from identity theft. This is the process of limiting the scope of the study to a particular clearly defined research problem and its specific elements.

One more vitally important point: At this stage, it is already necessary to use academic referencing style effectively, to cite all the works and sources being used. While there are a number of reference styles, and a particular course or university will advise which referencing style is required, we refer you to the APA seventh edition style guide, available at https://apastyle.apa.org/. Here you will find popular style guidelines, including title page setup, how to format your paper, bias-free language, in-text citations and reference examples. There are also webinars and tutorials freely available at https://apastyle.apa.org/instructional-aids/tutorials-webinars, for example, the webinar 'A step-by-step guide for APA student papers'.

4.3.3   Review of Scholarly Literature

A literature review is an important foundational step, which can help to build your knowledge foundations sufficiently to understand the problem domain, thereby arming the researcher with the ability to frame the research question clearly and concisely. Symbolically, this characterizes a narrowing step in the funnel stem. Thus, a researcher surveys scholarly papers including books, journals and conference proceedings relating to the subject area, compiles a descriptive summary of important ideas and themes and provides a critical review of the relevant works.

It is the expectation of the supervisor, the university, the principal investigator and fellow researchers that a researcher should spend ample time reading current journal articles on the subject area, as a way of igniting or directing the thinking and sense-making process. Figure 4.2 illustrates this initial phase of the research

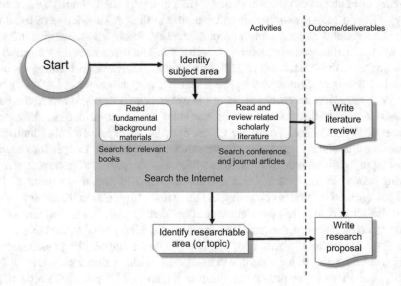

Figure 4.2 *Initial Phase of the Research Process*

process, demonstrating the steps that a researcher needs to undertake and the anticipated outcomes. As can be construed from the diagram, these are key activities that the researcher is expected to carry out within the wide mouth of the funnel, narrowing to the stem, in a logical sequence of activities, until the researcher has constructed a researchable topic.

Some researchers have argued that finding a researchable topic is not trivial, but a thorough literature review is capable of impelling the path towards a researchable topic (Pajares, 2007). This highlights the importance of identifying the main topic and associated keywords for the research, based on knowledge already gained from the background component of the study. The thorough literature review will help a researcher to understand the larger context of the study and knowledge already established about the topic. Conducting an extensive review of the literature in specific fields and exploring the references cited in those articles, book chapters and conference papers are an excellent starting point to your own research design.

At this stage, it is vital to think about the originality of your work and the possible contribution to knowledge. It is important to ensure that substantive work has not been done on the identical topic or alternatively that there is sufficient rationale to carry out further studies on the topic. It is also central to this part of the scoping process to ensure that there are sufficient sources of information to help the investigator substantiate the topic area. The literature review is discussed extensively in this chapter.

4.3.4   Identifying Gaps, Flaws, Defects and Weaknesses

The thoroughness of the literature review is not intended only to identify strengths in relevant works in the field but to find gaps, flaws, defects and weaknesses in existing studies. For example, the GDI (global diffusion of the Internet) framework was published in 2001 (Wolcott et al., 2001), at a time when public Internet use was in its infancy and when broadband services and mobile applications were virtually nonexistent. This requires a redesign of the framework, adapting the dimensions and levels to the digital environment of the 2020s, conducted by Mosina (2020). Identifying gaps, flaws and weaknesses demonstrates that the researcher has explored the existing knowledge to show where and how the proposed research fits into the field of study. It is expected that the researcher should substantiate how the literature contributes to the understanding of the problem domain under enquiry, by pointing to the strengths and gaps in the literature. The insights gained from the literature review also help the researcher to deal with conflicting concepts, resolve confusing issues and prevent duplication of effort, while justifying the necessity for further study. A detailed critical, analytical summation of relevant literature enables the researcher to scope the study and frame the research problem statement.

4.3.5   More Thoughts on Scoping the Research

The notion of scope brings the researcher to a vital decision point in a research study. It is concerned with establishing the boundaries of the research (what is included, what is excluded), so that the researcher can address the hypothesis or the main research question within set parameters. The objective is to clearly understand and precisely describe the limits of the study. Following the example in Figure 4.1, where the subject area is cybersecurity, it is noted that the field of cybersecurity is so vast that there is a need to narrow the topic to, say, cyber risk management. In this case, the researcher would focus on risk-related concepts and frameworks, not on cybersecurity in general. Thus, only background information, literature, theories and concepts relevant to cyber risk management will be included in the research problem statement, in the theoretical background and in the literature review. Thus, scoping is the product of the initial review of the existing body of knowledge in the general area of the field of study, that is, cybersecurity. The result of this scoping exercise is written up in the research problem statement, which states the main research problem and the three or four elements of that problem to be studied.

4.3.5.1   The Research Problem Statement

In academic research, writing a problem statement serves to improve the understanding and contextualization of the problem space and its significance. A well-understood problem is vital to resolving the problem and provides an effective solution (Pajares, 2007; Riggins & Wamba, 2015). In writing the statement, the research problem should be put in context, concisely, relevant to the problem domain, in such a way that offers the prospect of finding an answer, or practice-oriented response or solution or contribution to theoretical knowledge. It is important to understand that we need theory in order to build lasting solutions and innovations in any field.

4.3.5.2   Sample of Research Problem Statement

Over the past decade, financial institutions have sought to understand the factors contributing to large-scale cyber-attacks. The identity theft trend in the financial industry suggests that insiders' abuse of personally identifiable information (PII) accounts for many security breaches leading to identity theft and compromise of personal information. Identity theft accounts for huge financial losses in the banking industry which consequentially erodes customer trust. It has been argued that lack of effective cybersecurity governance, efficient legislation and regulation contributes significantly to identity theft in the financial services industry.

The above statement suggests that a quantitative or qualitative approach can help determine how the various factors of governance, legislation and regulation can affect identity theft. However, in typical quantitative research, the hypothesis can provide a statement to be tested using quantitative methods.

4.4   Design Project Versus Research Project

Not every project is an evidence-based research study. It is important to understand the difference between a design project and a research project (Hevner et al., 2004; McLaren & Buijs, 2011). The difference between the two is the underlying approach and process. In the digital sphere, one obvious characteristic of a design project is the adoption of an already established software development life cycle (SDLC) approach, without any attempt to refine the methodology. On the other hand, the research project will have a philosophical underpinning and apply investigative methodologies to probe the unknowns of a particular phenomenon (Hevner et al., 2004; McLaren & Buijs, 2011). Furthermore, a research study goes through the processes of data collection, data analysis, argumentation, explanation, verification and validation and may include implementation and testing (Oates, 2006). Thus, the research project is concerned with establishing new knowledge, whereas the design project is purely concerned with solving a particular problem without regard to new knowledge. While the research project is an academic work, which aims to contribute to the body of knowledge, in contrast, the design project seeks to solve a problem purely with existing knowledge. However, the outcome of academic research work may include the production of digital products or artefacts; thus CS (computer science), IS and CY research can include design creation methodologies (Oates, 2006).

The point here is that in some instances, research in CS, IS and CY is not satisfied with only the theoretical findings of the research study but requires the construction of a formal proof of concept, or prototype, to demonstrate the applicability of empirical findings, by using an established design and sound engineering principles. The idea is for the researcher to create a valid, repeatable scientific process to demonstrate that the outcome of research provides a solution to the real-world problem. In other words, a research study in CS, IS and CY can apply design principles to create proof of concept, based on the initial research findings.

For any computer science-based research output, the researcher is expected to demonstrate that the work is not a standard design or copy. The researcher must demonstrate those academic skills relating to applicability of methodologies, analysis, explanation, justification, argumentation and critical appraisal, alongside the attendant creation of new knowledge, not just demonstrating technical skills. An example of this is using SEMAT Essence Kernel and IBM Garage methodologies to analyse the strengths and weaknesses of the design process for an application for pest surveillance in agriculture, in order to improve the future app design (for further information on SEMAT [Software Engineering Method and Theory] see www. semat.org). Novelty and risk-taking are the hallmarks of research endeavours. Thus, researchers and students must distinguish the research project from the industry-oriented design project, noting that this should be done from the outset.

4.5   Developing the Flow of Ideas

The funnel strategy diagram depicts the mental process that generates the flow of ideas that should be followed in the design of a research study, without setting out the details of the tasks themselves. Many tasks in research work will require the researcher to develop a flow diagram, guiding the research process. Figure 4.3 illustrates a typical flow of ideas to develop a research plan and includes the activities that are required at each step.

The first step is to narrow down the range of possibilities to the particular topic of interest. The researcher must understand the motivation and state it clearly and concisely. With this in mind, the researcher seeks to develop a statement that describes the particular research problem and the reason why addressing that problem is imperative. Confirming that the problem exists requires a concerted effort, through the review of relevant background information available in reports and other documents, as well as the most relevant scholarly literature in the field. By navigating from one set of documents to another, and from one set of thematic articles to another, the researcher can identify and discuss the related arguments or positions, thus establishing the evidence for the nature and characteristics of the research problem. Weaknesses, defects, flaws, shortcomings and suggestions for future works in those articles can guide the researcher to the focus for your particular study.

The development of ideas can be stimulated by certain questions that the researcher poses to her/himself, such as the following:

- Where does the problem lie? Who does the problem affect? What is the context?
- What is the overarching research problem?
- What are the three key related elements of this research problem?
- How will it affect the economic sector, or social sector, if the findings are able to address the problem?
- Is there a timeline to fix the problem?

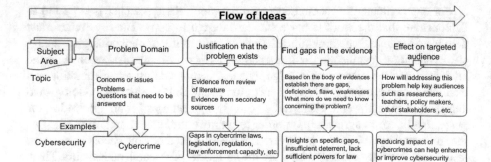

Figure 4.3   *Typical Research Task Flow*

- Is it all important to fix the problem?
- Is the problem a temporary issue or is there a likelihood that it continues into the future?
- Is the research capable of revising existing knowledge or practices? If so, how?

Once the answers to these questions are provided, the researcher should have a well-rounded problem statement and make a few drafts until the problem statement is as polished as possible.

Therefore, we can conclude that the funnel strategy stimulates the flow of ideas in the early phase of research design in a logical and structured manner, and using this approach can be a valuable asset to the researcher. In CS, IS and CY research, it is easy to be side-tracked by the sea of information that is accessible. By using the funnel strategy, the researcher can direct herself/himself to remain focused. This will allow the researcher to achieve better outcomes and avoid wasting time and effort due to taking unnecessary detours away from the core goal.

4.6   Mind Mapping Applied to Research Design

4.6.1   The Concept of Mind Mapping

Contemporary research requires a well-structured sequence of logical and structured thinking, and a strong flow of ideas, to drive an acceptable scientific or social enquiry. This is challenging to many new researchers. Crafting research topics, perspectives and problem statements, from the apparent chaos of ideas in the literature into logical, structured and creative arrangements, combined with a flow of logical thought patterns, is necessary. The researcher can use mind mapping to derive aims, design objectives and write up the hypothesis (quantitative research) or problem statement (qualitative research). Mapping ideas enables us to organize ideas into coherent thought (Wilson et al., 2016). It helps to increase the flow of understanding and makes fresh connections highly visible, to create many possibilities of arranging thoughts, ideas and relationships in a logical order. The technique fosters creativity by enabling intuitive thinking that builds on existing knowledge. It combines appropriate logical and structured thinking, with intuitive thinking. Researchers can derive hypothesis, problem statements, objectives, questions, approaches and methods and explore validation through mind mapping. Usually, visual organization of ideas is based on hierarchical order showing the appropriate relationship among related concepts (Crowe & Sheppard, 2016); however, mind mapping approaches and tools enable the researcher to draw many distinct kinds of relationships, crisscrossing themes and categories and showing dynamic linkages across a broad set of ideas. A sub-topic, concept or theme can be mapped onto the main topic, which allows integration of several ideas in a nonlinear, graphical layout that enables the building of an intuitive framework around a central concept (ThinkBuzan, 2010).

Mind maps are graphical representations of data, information and thoughts, as opposed to traditional linear notes or annotations in text documents. Consequently, mind mapping can aid the researcher to capture thoughts, ideas and keywords and organize them in a multidimensional structure. The central idea or title is always placed at the centre of the mind map, and related ideas branch off from the centre in multiple directions, thereby establishing a radiating structure. A mind map can be accomplished with manual drawing, using rough notes from meetings or technical sessions, and there are also a range of automated software tools that can aid in mind mapping (see, e.g., Software Testing Help at https://www.softwaretestinghelp.com/mind-map-software/). With mind mapping, it becomes easier to brainstorm and then organize thoughts into structure and order. It assists you to visually structure your ideas for synthesis, analysis and recall.

4.6.2   Mind Map Use Cases and Benefits

Mind mapping techniques can be applied to any research process, converting unstructured information into a memorable and organized structure. It enables the researcher to visualize the bigger picture and to build connections and hierarchies of ideas. It activates the brain and enhances the thinking process by applying mental triggers. In addition, mind mapping empowers unrestricted flow of thoughts and enhances memory. By using mostly single keywords, symbols and phrases in creating mind maps, the researcher can jot down thoughts in a faster way, which encourages an unhindered flow of ideas.

Mind maps are not meant only for applications in research but can be applied to many diverse tasks, including lecture note taking, as a study aid, brainstorming individually or in groups, problem-solving, gaining insight into complex subjects and research project planning. By so doing, research productivity is increased in a time-saving manner. Figure 4.4 shows a typical mind mapping exercise in the academic research context. It is very useful in writing research proposals since it provides researchers with a logical way of thinking, as well as building the timeframe for the research design.

4.6.3   Application of Mind Mapping to Setting Out the Research Tasks

Using mind mapping to draft the outline of your research proposal and plan the layout of your thesis can be a significant contributor to achieving success. The mind map guides the entire research process, allowing systematic flow in the research approach. When applied to setting out the tasks for the research proposal,

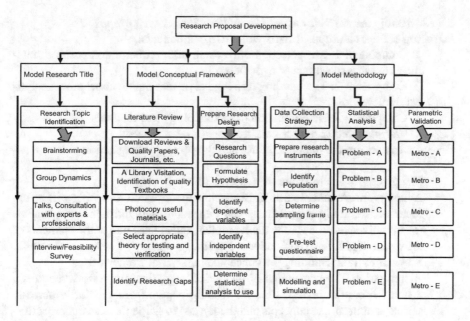

Figure 4.4   *Mind Mapping Framework for Research Models/Proposals*

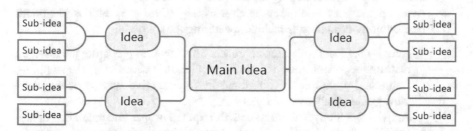

Figure 4.5   *Idea Synthesis in Mind Map*

it allows you to plan extensively, providing critical details that need to be addressed, within a set time frame, followed by monitoring the planned progress. The mind map illustrated in Figure 4.5 presents a mind map format for your ideas or research tasks.

In order to create a mind map for writing the research proposal, the researcher should consider the following steps:

Step 1: Identify the main goal.
   Since the intention is to prepare the research proposal, the researcher's goal will be highlighted as *Prepare Research Proposal*: Type this central idea in the 'main idea' box.

Step 2: Identify the activities needed to prepare the research proposal.
Preparing a research proposal starts with three components:

    1. Component 1: The selection of research title, problem and purpose statement research questions or hypotheses

    2. Component 2: The building of a logical or reasonable conceptual/analytical framework

    3. Component 3: Determination of methodology and methods

Step 3: Determine and specify the tasks that must be achieved.
Here you can set out the major activities and the corresponding action points:

Major activity 1: Preparing the research title
Action point: A suitable title must identify the research topic that will be the focus of the study, highlighting the dependent and independent variables (quantitative research) or the main problem and at least one of the elements of the research problem (qualitative research).

Major activity 2: Building the conceptual/analytical framework
Action point: The conceptual or theoretical analytical framework requires a detailed literature review, time for reading papers from scholarly journals, selecting appropriate concepts and theory(ies) to guide the research, identifying gaps in knowledge to be addressed, drawing an analytical framework diagram and including a short explanation of the diagram.

Major activity 3: Developing the research methodology
Action point: Based on the research gaps, objectives or questions identified, methodology design tasks include specifying the following:

Activity 3.1 Data collection approach: This involves preparing the research instruments, identifying the study population, determining the sampling frame and pretesting the questionnaire (quantitative survey) or the interview guide (qualitative study).

Activity 3.2 Statistical analysis: This requires proper understanding of statistical tools and how to carry oust statistical analysis for each hypothesis, where the study is quantitative. Parametric validations are also vital to validate the researcher's expected/proposed result.

These steps can be modified while assigning timeframes for each of the activities that you have identified. In so doing, the researcher will be able to make a rough estimate of how long it will take to be able to prepare the research proposal. If the researcher has a fixed period within which to submit the draft proposal, the researcher can allocate specific periods for the accomplishment of each activity, which can facilitate submitting the proposal on time. Figure 4.6 gives a complete picture of the mind map applied to writing the research proposal. Leveraging this mind mapping approach will produce a good quality research proposal, noting that you also need to pay attention to the quality of the content.

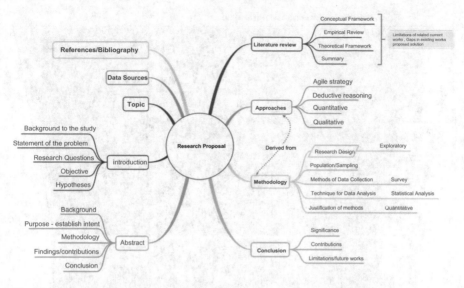

Figure 4.6 *Mind Map for Research Proposals*

4.7 Applying the Funnel Strategy and Mind Mapping to the Research Content

In this section, an attempt is made to demonstrate how the funnel strategy and mind mapping can stir a researcher in conceptualizing and conducting a research study. The thrust of these two tools is to direct and aid the flow of thoughts, which can drive the concept mapping of the research parameters and connects existing pieces of evidence to the fresh ideas developed. So we begin with the central idea of the subject area or topic and work through to situate and address the problem in the context of the research constraints. The following diagram explains how to apply the tools described in this chapter.

4.7.1 Example 4.7.1: Autonomous Agent and Multiagent Systems

Figure 4.7 illustrates how the funnel strategy is applied to conduct research in the area of autonomous agent and multiagent systems.

Having identified the broad topic of interest through learning about the field and conducting an initial scan of journal articles, the next step is to engage in extensive background reading of established concepts and theories relating to the topic. For instance, most complex intelligent software systems require a good knowledge of algorithms, theories and assumptions, so learning from existing ones, and gaining

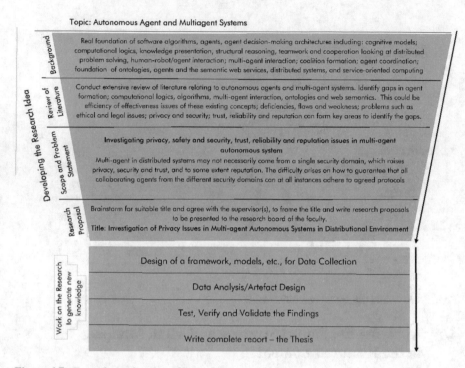

Figure 4.7 *Example Application of Funnel Strategy*

an understanding of how to create fresh algorithms should help build the researcher's intellectual capacity. Software agents are built based on sound decision-making, cogent reasoning, computational logic and communication protocols. It is important to engage in extensive background reading with respect to the key foundational concepts relating to the topic, to sharpen the researcher's knowledge and understanding, before reviewing the literature in depth. The knowledge gained during the extensive reading on foundational concepts will help the researcher to understand and present in writing these concepts found in literature, in your own words. Figure 4.8 depicts how to apply mind mapping as a way to do concept mapping on the research topic. Radiating out from the centre text box, the first-order elements randomly show important sub-topics, and subsequently, branches are created to reflect the lucid thinking of the researcher with respect to the concepts, important variables and parameters relevant to each sub-topic. The supposition is that if a researcher can carefully and systematically learn from previous research works, she/he can comfortably frame and construct a new research enquiry. When fresh thinking is interconnected with an existing body of knowledge, then that fresh thinking is a lot easier to present, explain and write up in the text of the research proposal. So the researcher has the responsibility to create a scaffolding of ideas from an existing body of knowledge, in order to introduce fresh knowledge. Doing concept mapping is one way to scaffold ideas and design research parameters (what is included in your study, what is excluded from your study), also referred to as defining the scope or boundaries of the study.

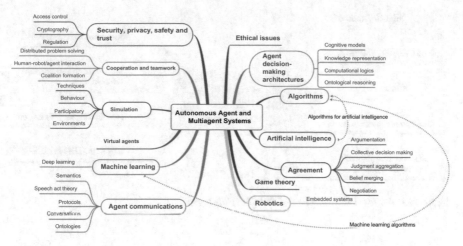

Figure 4.8 *Concept Mapping of Autonomous Agent and Multiagent Systems*

4.7.2   Example 4.7.2: Artificial Intelligence (AI) and Robotics

Similar to the example discussed in Section 4.6.1, a research enquiry in the subject area of artificial intelligence (AI) and robotics can cut across several disciplines, requiring the researcher to contextualize the investigation in a multidisciplinary or alternatively in an interdisciplinary way. Figure 4.9 depicts the topic of artificial intelligence and robotics, narrowing to the possible domains or sub-topics within this subject area.

What this entails is that the researcher will focus on the conceptual and theoretical frameworks that deepen the understanding of background information on any of the sub-topics. However, the researcher must gain a thorough understanding of the basic AI and robotics subject area, before diving deeper into the context of the sub-topics. The researcher must seek to understand the present 'state of the art' and imminent research needs and to identify important research gaps, through connecting the background information, concepts, theories and assumptions underlying the study, to the discussion of those concepts and theories in scholarly articles. The researcher must understand the foundational computer science concepts needed to enable her/him to conduct research in AI and robotics, notwithstanding which particular direction the research will take. Applying the funnel strategy, background reading is the first step, after the researcher has chosen the sub-topic(s).

Figure 4.10 demonstrates the fundamental constructs that can underpin the understanding of AI and robotics and further learning. For background reading, the researcher must comprehend such concepts as neural networks, machine learning, data mining, deep learning, big data, intelligent systems, algorithms, kinematic transformations, reverse kinematics, etc., regardless of the area of concentration.

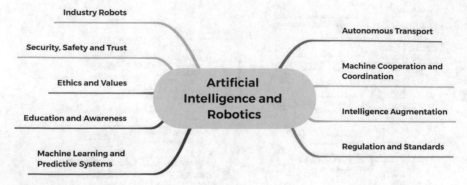

Figure 4.9 *Artificial Intelligence and Robotics Sub-topics*

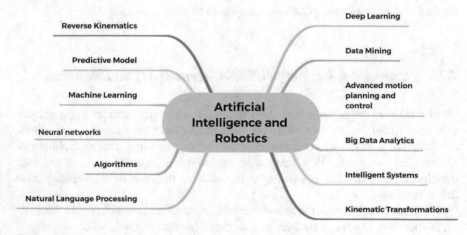

Figure 4.10 *Foundational Constructs in AI and Robotics*

Generating a clear understanding of these concepts in the proposal will assist in building a deeper understanding of the topic and sub-topics, creating the path to situating and scoping the study. Reading as widely as possible and understanding relevant theories and concepts is necessary to writing a good literature review, where the most important ideas are discussed, and the research gap is clearly identified. Having scanned and read widely, the next step is to select those journal articles and peer-reviewed conference papers, theses and dissertations that will be used for the literature review, all guided by the decision on the scoping of the topic and sub-topic. For example, the selected sub-topic could be ethics and values. In this case, the researcher can narrow the scope of the study to identifying the gaps in research in ethics and values. The next step is to brainstorm contextually and frame the

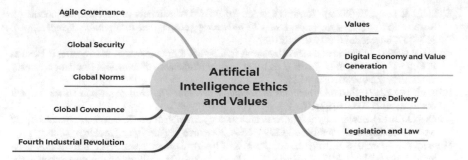

Figure 4.11 *Concept Mapping Extension of Ethics and Values*

problem to be addressed and decide on a suitable research title. For instance, such research title can be framed as *investigation of ethics and values in autonomous transport systems*. The next step is to develop the research proposal to reflect the chosen topic.

At this point, the researcher has arrived at the stem of the funnel, which metaphorically illustrates that the researcher has finally found the direction of the study, and henceforth can sustain and remain focused until the research is completed. In the case of this example, we now use a mind map to draw the underlying concepts and parameters to suit the specific scope of the study. Figure 4.11 illustrates a conceptual perspective of AI ethics and values that the researcher can apply to the specific context of autonomous transport systems.

The mind maps shown in Figure 4.5 through to Figure 4.11 show the cascade of ideas that is possible through the use of mind mapping tools and techniques. With this sort of mapping, the researcher is armed with an array of possible alternatives to narrow and scope the study, in such a way that a fresh contribution to the body of knowledge is achievable. The point here is that the researcher must explore the literature broadly, while at the same time bearing in mind the constraints of the study.

Bibliography[1]

APA. (n.d.). *APA style guide*. https://apastyle.apa.org/
Carlsson, S. A. (2006). Design science research in information systems: A critical realist perspective. In *ACIS 2006 Proceedings, Australia*. https://aisel.aisnet.org/acis2006/40/
Crowe, M., & Sheppard, L. (2016). Mind mapping research methods. *Quality & Quantity, 46*, 1493–1504. https://doi.org/10.1007/s11135-011-9463-8

---

[1] If you find that any of the following material is not downloadable at the listed url, please search on ResearchGate at https://www.researchgate.net/ or, alternatively, search on the full name of the article, book, or paper.

Ellis, T., & Levy, Y. (2009). Towards a guide for novice researchers on research methodology: Review and proposed methods. *Issues in Informing Science and Information Technology, 6*, 323–337. https://doi.org/10.28945/1062

Ernest, P. (1996). Varieties of constructivism: A framework for comparison. In L. Steffe, P. Nesher, P. Cobb, G. Goldin, & B. Greer (Eds.), *Theories of mathematical learning*. Routledge. https://doi.org/10.4324/9780203053126

Hassani, H. (2017). *Research methods in computer science: The challenges and issues*. https://arxiv.org/pdf/1703.04080.pdf

Hauberg,J.(2011).Researchbydesign:Aresearchstrategy.*Architecture & Education Journal,5*.https://www.researchgate.net/publication/279466514_Research_by_design_a_research_strategy

Hevner, A., March, S., Park, J., & Ram, S. (2004). Design science in information systems research. *MIS Quarterly, 28*(1), 75–105. https://www.researchgate.net/publication/201168946_Design_Science_in_Information_Systems_Research

Indulska, M., & Recker, J. (2010). Design science in IS research: A literature analysis. In S. Gregor & D. Hart (Eds.), *Information systems foundations: The role of design science*. ANU Press. https://doi.org/10.22459/isf.12.2010.13

MacIntyre, S., Dalkir, K., Paul, P., & Kitimbo, I. (2014). *Utilizing evidence-based lessons learned for enhanced organizational innovation and change*. Business Science Reference.

McLaren, T., & Buijs, P. (2011). A design science approach for developing information systems research instruments. In H. Jain, A. Sinha & P. Vitharana (Eds.), *Lecture notes in computer science: Vol. 6629. Service-oriented perspectives in design science research* (pp. 1–10). Springer. http://www.rug.nl/staff/p.buijs/design_science_approach_for_developing_isr_instruments.pdf

Mosina, C. (2020). *Understanding the diffusion of the Internet: Redesigning the global diffusion of the Internet framework*. [Research report, Master of Arts in ICT Policy and Regulation]. LINK Centre, University of the Witwatersrand. http://wiredspace.wits.ac.za/handle/10539/30723

Oates, B. J. (2006). *Researching information systems and computing*. SAGE Publications. https://books.google.co.za/books/about/Researching_Information_Systems_and_Comp.html?id=ztrj8aph-4sC

Pajares, F. (2007). *The elements of a proposal*. Emory University. https://www.uky.edu/~eushe2/Pajares/proposal.html

Riggins, F. J., & Wamba, S. F. (2015). Research directions on the adoption, usage, and impact of the internet of things through the use of big data analytics. Proceedings of the *2015 48th Hawaii International Conference on Systems Science, USA*, 1531–1540. https://doi.org/10.1109/HICSS.2015.186

Salhin, A., Kyiu, A., Taheri, B., Porter, C., Valantasis-Kanellos, N., & König, C. (2016). Quantitative data gathering methods and techniques. In A. Paterson, D. Leung, W. Jackson, R. MacIntosh, & K. D. O'Gorman (Eds.), *Research methods for accounting and finance*. Goodfellow Publishers. https://doi.org/10.23912/978-1-910158-88-3-3226

Software Testing Help. (2021, March 27). *10 best mind mapping software in 2021*. [website]. https://www.softwaretestinghelp.com/mind-map-software/

ThinkBuzan Ltd. (2010). *Mind mapping: Scientific research and studies*. [online]. https://www.slideshare.net/elsavonlicy/mind-mapping-evidence-report

Wilson, K., Copeland-Solas, E., & Guthrie-Dixon, N. (2016). A preliminary study on the use of mind mapping as a visual-learning strategy, in general education science classes for Arabic speakers in The United Arab Emirates. *Journal of the Scholarship of Teaching and Learning, 16*(1), 31–52. https://doi.org/10.14434/josotl.v16i1.19181

Wolcott, P., Press, L., McHenry, W., Goodman, S., & Foster, W. (2001). A framework for assessing the global diffusion of the Internet. *Journal of the Association for Information Systems, 2*. https://doi.org/10.17705/1jais.00018

Chapter 5
Foundational Research Writing, Background Discussion and Literature Review for CS, IS and CY

5.1 Introduction

Before getting to the practical task of data collection, it is important to evaluate the foundational components for undertaking the research endeavour, namely, the background discussion and the literature review. In this chapter, we discuss their content and their respective roles and significance in the CS (computer science), IS (information systems) and CY (cybersecurity) research study. We need to know how the foundational research writing, the background discussion section and the literature review fit into the full range of research activities and how they enable the researcher to complete the study. Foundational research skills include vital processes, from framing a statement of the research problem to designing the analytical framework.

This chapter is dedicated to explaining the skills a researcher needs, the type of background information required to support the research enquiry and the literature review, which helps to situate the research in context, and lays the foundation for verification and validation of the research work. Learning research skills and techniques builds and strengthens the capacity to conduct research. Research capacity requires the ability to identify and describe/present a problem, determine the research purpose, outline objectives and set priorities for the study. It creates the mental disposition for the researcher to engage in scientific and empirical research and identify critical paths to understanding and/or solving the problem.

Consequently, when embarking on postgraduate study, the researcher must concentrate on building and strengthening her/his research skills, in order to ensure that the research outputs will be valuable to particular communities of interest, whether these communities are academic or organizational or practitioner communities in the relevant field. Research capacity helps a researcher, individually and collaboratively, to undertake research effectively and proficiently and in a sustainable manner. Furthermore, background reading in the subject area helps the researcher to understand the broad theories and concepts that underpin the field of study. The literature review situates the research in the context of existing literature, to position how the study relates to the existing body of knowledge and to identify strengths and establish gaps that exist in published work. The literature review can aid the researcher to provide a comparative synthesis of related works, resolve conflicts in contradictory opinions, ensure that the research study has a high degree of relevance and ensure that the researcher avoids unnecessary duplication of already published research.

In this chapter, we take the reader through the foundational skills needed to build the capacity for conducting a research study. We discuss the relationship between the research problem statement, the background information required (practice-based, theoretical and conceptual) and the design of the literature review and analytical framework.

5.2   Foundational Research Tools and Techniques

The research design exercise should commence with writing a draft of the research problem statement (about 200–300 words) as the first task. A simple way to think about a research problem statement is a statement that indicates to the reader what is not currently known, or published, about the topic under discussion. The statement does not contain any questions, nor does it need to contain any references, as it is a statement of **your** understanding of the research problem that needs to be investigated. The research problem is selected, because the new data and analysis will reveal new insights, visualizations, responses, options, opportunities and/or solutions. The research problem statement should include an indication of the main research problem (e.g. cybersecurity vulnerability in banks) and at least three, but preferably not more than four, themes relevant to the research problem. This creates the scope for the study.

The research problem statement is followed by the research purpose statement, the main research question, the three or four research sub-questions and the background discussion. This constitutes the first major section of the research proposal and becomes the first chapter of the final report. The second major section of the report is the literature review. The difference between the background discussion and the literature review is as follows:

Difference 1: The background discussion provides existing data about what is known about the research problem in its context. This can include the number of banks and banking customers in the study country, the number and types of cybersecurity incidents experienced in that country in the past three to five years, the increase or decrease in cybersecurity incidents and financial losses over time.

Difference 2: The literature review considers the concepts, themes and theories apparent in the scholarly literature, in particular in published journal articles and their relevance to the research problem. It confirms the gap in knowledge and lays the foundation for later data analysis.

5.2.1   Writing the Annotated Bibliography

The next task is to write the annotated bibliography with respect to all the readings that you will use, both the readings for the background discussion and the readings for the literature review. It is essential to make notes (annotations) as you are reading, to ensure that the reading has value to your research design. The annotated bibliography for the background discussion includes readings on trends pertaining to the research problem, for example, briefly sketching the trends in cybersecurity vulnerability in general and in the banking system in particular, in order to identify the three or four main themes relevant to investigating cybersecurity vulnerability. The annotated bibliography for the literature review relates to identifying how relevant theory and concepts have been applied, or are being applied, in relation to cybersecurity vulnerability. An annotated bibliography is an organized way of reading, extracting highlights and writing critical analytical commentaries, as the foundational notes that you will use to write the background section and the literature review section of the research paper. We discuss reading and critical analysis in more detail below.

A few words of guidance on the steps for writing an annotated bibliography are as follows:

Step 1:   Write the correct full APA (American Psychological Association) style reference (or another reference system that you are required to use). The APA offers a free online APA seventh edition style guide, tutorial and blog available at https://apastyle.apa.org/. You can use the 'Popular Style Guidelines' (Reference Examples) to see how to reference journal articles, books and e-books, reports with individual authors, conference proceedings and other scholarly literature. The Harvard Format Citation Guide is available at https://www.mendeley.com/guides/harvard-citation-guide. You can find other reference style guides available from reputable university sources on the Internet.

Step 2:   Highlight of key points from the article: Think carefully about the meaning of highlights – these should be the most powerful ideas from the article or book chapter, usually not more than two or three ideas.

Step 3:   Analytical commentary on the article: This refers to presenting your own analysis of the highlights of the article, which would include the strengths and weaknesses of the ideas, and their relevance to your research problem statement, the type of organization or sector, the specific country or all of these. It can also include an indication of what you see as gaps, serious flaws or limitations and possible alternative views.

Step 4:   The annotated bibliography should relate the highlights and analytical comments on each article to the specifics of the research problem statement.

The next task is to write the background using the information from the annotated bibliography. The following task is to write the literature review of the articles, noting their relation to each other. When preparing the literature review section, build your own argument about how the literature you use sheds light on the issue(s) raised in the problem statement.

5.2.2   *Reading and Writing with Purpose*

All stages of scholarly research work involves extensive reading and understanding. A researcher should develop a sustainable appetite to read as widely as possible, at the early stage of learning research skills. A researcher must read to understand and ascribe meaning to what is read. The development of reading prowess will enable the researcher to have the aptitude to assimilate ideas easily, within a short space of time. But sometimes, the researcher is overwhelmed by the vast amount of information available and does not know how to navigate through this information systematically, to gain insight from the content. Figure 5.1 presents a pragmatic and simple flow chart, which can guide the researcher on how to read and write effectively in a workable and productive manner.

Once the topic for the research has been decided, the researcher should write down the most important search terms including the main term and the terms most closely related to the main term, for example cybersecurity vulnerability (main term), network security, applications security, internal vulnerabilities, external vulnerabilities and human vulnerabilities (related terms). Select several readings from published reports and books that include these search terms (between 12 and 24 readings from the study context, including industry, or government, or economy or society trends, as well as from theory) – this will create a foundation for writing the background to the research problem statement. In addition, select several journal articles that include these search terms (between 12 and 24 journal articles) – this

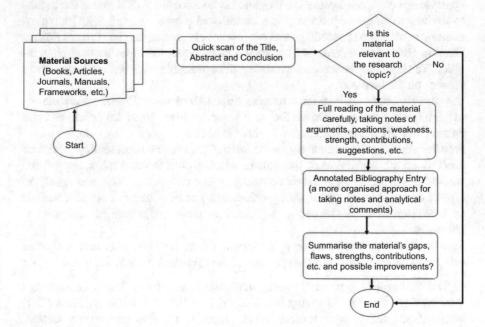

Figure 5.1 *Background and Literature Review Flow Chart*

will create a foundation for writing the literature review relevant to the research problem statement. Scan through the articles, following the steps in the chart below. This initial reading process is vital in selecting articles that are closely related to the purpose of the research. Focus on establishing whether the topic is researchable and gauge the nature of any ongoing research work in the area. In general, Figure 5.1 should guide the selection and reading of relevant materials, in order to maximize time and productivity.

The researcher must be able to decode the relevant concepts/terminology and understand their meanings, evaluate the applicability/value of the article to the research and decide which readings and articles to retain. Academic research demands significant attention to reading, critical thinking, analytical skills and the capacity to do all of these, implying that a researcher must spend ample time reading, thinking and analysing. Adapting to this systematic and constructive reading approach enables the researcher to build a rich vocabulary, concepts and taxonomy of the topic. The annotated bibliography is a key step in the process of translating reading into academic writing.

5.2.3   Critical Thinking and Analysis

Critical thinking applies to many areas of development, not only to academic research. It is explained in different ways by different authors. Critical thinking includes comparative review, reflective thinking, independent thinking, deductive reasoning, inductive reasoning and counter-inductive reasoning. A researcher needs to do comparative analysis of ideas from diverse sources and identify their logical connections to other ideas. Reflective thinking means that you can compare and contrast the ideas from authors with your own experience of the particular issues under discussion. Independent thinking refers to a state of mind where the researcher challenges existing views presented by authors and considers possible alternative ways of understanding the issues under discussion. Deductive reasoning applies when we use the literature to create a hypothesis of what will occur, based on our understanding of theory, and then (later in the research process) make conclusions about whether the data collected fits the hypothesis or does not fit the hypothesis. In deductive reasoning, we move from the general (theory about X) to the specific (does the theory apply or not, how does it apply). In inductive reasoning, we search for and identify patterns, sequences, categories and themes in the literature and then build an analytical framework based on analysis of the literature. This analytical framework can then be applied to thinking about what the data means. Counter-inductive reasoning means exploring alternative possibilities, alternative possible realities or alternative options to the obvious ideas presented in the literature. In the process of critical analysis of existing literature, the researcher needs the ability to explain concepts, evaluate ideas, reflect and draw inferences from arguments, observe events or phenomena, interpret and communicate events or ideas, analyse evolving situations, coherently organize ideas and be creative and adopt open-mindedness and problem-solving attitudes. As discussed above, examples of critical analysis include the following:

Example 1: Revealing and documenting the strengths and weaknesses of the ideas of various authors

Example 2: Explaining how these ideas are relevant to the research problem statement

Example 3: Explaining how the ideas are relevant to the organizational, social, economic or country context of the study

From a research perspective, critical thinking is the use of logical and structured reasoning to be an active contributor to knowledge, rather than an inert beneficiary of ideas. Critical thinkers can meticulously interrogate ideas and norms, as opposed to simply absorbing them without questioning. Critical thinking requires a researcher to clarify whether the theories, concepts, ideas, arguments and results characterize the holistic view of the subject or are open to extension. Figure 5.2 shows the complementary aptitudes a researcher needs to have to develop critical thinking, and all these elements are interlinked and can enable writing a productive research study.

Every research study needs the researcher to make important decisions, and the outcome of research depends on the decisions made at each stage of the enquiry. Specifically, for sound decision-making, the researcher needs to think thoroughly about the problem space in a rational, structured and critical manner, amidst a range of alternative perspectives; the researcher needs to interrogate the diverse arguments and positions concerning the particular issue, debate or phenomenon. The researcher

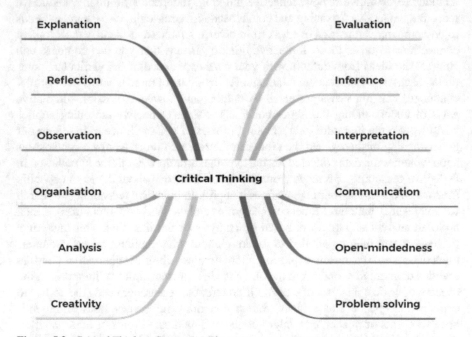

Figure 5.2 *Critical Thinking Supportive Elements*

needs to appraise several viewpoints and arguments presented in the literature, in order to ascertain their veracity, identify flaws, drawbacks, weaknesses or inappropriate positions of argument and must limit her/his own prejudices. The ability to avoid the inclination to take a particular side in the debate, or adopt an opinion, without thorough analysis, will reinforce good research outcomes. Critical thinking helps to provide structured reasoning and support for the position taken in a well-argued analysis. So the critical thinking ability of a researcher must be combined with the capability to observe objectively, explain a phenomenon carefully, organize thoughts and ideas logically, analyse competing arguments, be open-minded to accept novel positions, interpret events, deduce inferences and communicate new positions to others.

To sum up, when a researcher develops critical thinking skills, research decisions will be evidence-based, which can invariably lead to achieving the best conceivable outcomes. Note that research involves gathering, examining and synthesizing information from a variety of sources; hence critical thinking aids in revising the outcomes and applying necessary changes where desirable.

5.3   Background Discussion

The initial stage of research enquiry requires extensive reading to familiarize the researcher with the underlying concepts and theories behind the problem of interest. Thorough background study of the key concepts stimulates and motivates the researcher and can facilitate concept mapping of basic ideas, once the researcher ascertains a topic of interest. For instance, if the researcher is interested in modelling an aircraft's weight density, background reading will focus on fundamental concepts associated with aerodynamics, kinematics and rigid bodies and dynamics; and building thoughts around these core concepts can help interlink several of these insights into a logical understanding. If the researcher is interested in cybersecurity vulnerability, then he/she will focus on concepts associated with technological vulnerability (e.g. network security, applications security, security by design), business vulnerability (presence or absence of cybersecurity strategy, weak or strong cybersecurity management) and human vulnerability (e.g. insider threats, password protection, cybersecurity skills deficit).

The best approach to fuel a researcher's inspiration, especially when planning to develop researchable questions, is by extensive reading, scanning for ideas and looking at and narrowing of the topic of interest, as described in Chap. 3. The effort must be made to generate detailed ideas and to focus on connecting varying degrees of information, both structurally and logically. There are three possible components to the background section, each of which should be a reasonable word length. These are practice-oriented perspectives and trends, theoretical background and conceptual background.

5.3.1 Practice-Oriented Background and Review of Trends

An important component of the background to the research problem statement is a review of the current trends and the drivers of those trends, in other words, what is happening in the real world (practice orientation), for example, trends in cybersecurity vulnerability in a particular sector, such as banking and financial services, and what is happening in that same sector at national level, at African level and at global level. This background data can be obtained from the following sources:

Source 1: Industry reports, for example, organizations in the financial sector release annual reports on the state of the industry in a particular country, banking industry forecasts (Mensah et al., 2019) or specifically on the state of banking crime (SABRIC, 2019), and research consulting firms such as Serianu publish country-level reports.

Source 2: Global Trends reports about practice, for example, the International Telecommunication Union (ITU), the ISACA, the World Economic Forum and the World Bank, release many annual reports on economic and other trends. In particular, ISACA releases an annual report on cybersecurity trends and Serianu releases industry reports and annual reports on cybersecurity trends in African countries (Serianu, 2017, 2018, 2019, 2020a, b).

Source 3: Annual reports or strategy documents of firms and organizations, which are generally available on the Internet.

5.3.2 Theoretical Background

Theoretical assumptions and discussion are the cornerstone of enquiry into the research problem, framed to clarify and comprehend the phenomenon of study and/or to forecast evolving events. The aim is generally to interrogate, and to advance what is already published in the universe of knowledge, as a basis for new thinking. So we can say that the theoretical background will be the foundation for interrogating a particular research problem, in a particular field of study. The theoretical background presents the theories that explain why the research problem exists and/or what constitutes the research problem (as far as we know). As a vital component of a research enquiry, the initial concepts, assumptions or theories that have relevance to the research problem should be clearly stated in the background discussion. The idea is that contemporary studies are usually underpinned by theories that describe interrelated phenomena. Thus, theory offers the foundational concepts, assumptions and holistic theoretical framing that a researcher needs to be familiar with, in order to support the understanding of the research subject (Abend, 2008). Theory helps to situate the topic of research in the broader domain of knowledge being studied and enables the researcher to be more grounded, as the basis for producing meaningful results. The introduction of the relevant concepts and theories can be kept relatively short in the background discussion, as they can be discussed at greater length in the literature review section. Researchers often make the error of trying to draw the theoretical background from peer-reviewed journal articles and conference proceedings alone. It is important to emphasize that lectures, books, scientific research studies and observation of the phenomenon (research problem) as presented in media reports can often be the sources of ideas about theoretical assumptions, conceptual models and theories, which can then be explored in greater depth in the relevant scholarly literature.

Next, we show how to use mind mapping to conceptualize the theoretical background for the selected research topic area. For example, a researcher is developing a basis for understanding multifactor authentication to support a particular application domain or a generalized research enquiry. Figure 5.3 illustrates the idea of framing a theoretical background using a mind map. The figure illustrates that a researcher should understand the theories, assumptions and concepts relevant to multifactor authentication, by studying access control models, biometric models, network security, computer security, application security, etc., in order to be theoretically grounded in the area under enquiry. In the conceptual background section, we will draw a sharp contrast to help the researcher differentiate these two components of the background to the study.

5.3.3   Conceptual Background

A conceptual background is used to introduce (or plant) the foundational concepts, parameters and variables that may be relevant to the expected findings. The primary source of framing the conceptual background is usually relevant peer-reviewed journal articles or conference proceedings, especially the keywords and synthesis used to describe the dependent and independent variables relating to the study. The selection of appropriate concepts is usually based on the researcher's ability to draw inferences from the various arguments, perspectives or positions taken on the subject.

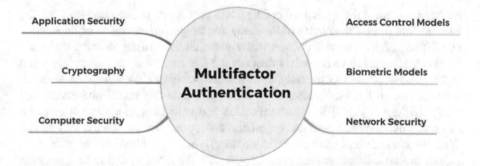

Figure 5.3 *Conceptualization of Theoretical Background for Multifactor Authentication Research*

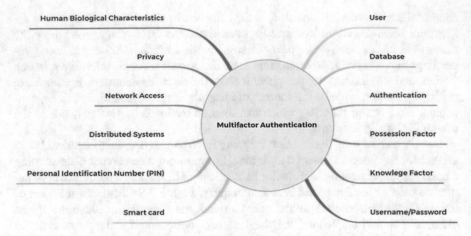

Figure 5.4 *Conceptual Framework Concept Mapping*

Formulating a conceptual background can be challenging, due to the fact that it is dependent upon the type of research. Where research has a cause-and-effect character, the researcher should aim to identify the variables that illustrate the relationships that exist among the dependent and independent variables. In our previous example, in relation to understanding multifactor authentication, Figure 5.4 depicts the mapping of elements of the conceptual background that may be relevant to the multifactor authentication research problem. The researcher can frame the dependent and independent variables that are germane to the research study depending on the scope. At a glance, it reflects the factors that may influence the input and output variables, but more importantly, it illustrates how they can firmly connect to theoretical constructs.

5.3.4   Theoretical Background Versus Conceptual Background

So far, it is evident that both the theoretical and conceptual backgrounds provide the impetus that directs the course of the study and drives the extension of knowledge in the area of the study. However, many researchers confuse the two elements. Research scholars have described them as the life that motivates empiricism and rigour (Imenda, 2014). Conversely, some research scholars have argued that research findings can be faulted because of their inappropriate theoretical and conceptual design (Adom et al., 2018). The verification and validation of a piece of research can suffer due to insufficient theoretical and conceptual design, since it may be difficult for scholars/readers to ascertain the underlying foundations of the study.

The theoretical background is concerned with the theories that initially explain the nature of the research phenomenon (Swanson, 2013), though existing theory may offer an incomplete picture. The theoretical background formulates a foundational thinking and analytical structure for the enquiry. On the other hand, a conceptual background is concerned with causal variables and/or concepts that underpin the area of study, especially empirical findings from previous enquiries. It is important to show the relationships among these concepts, as well as how they relate to the research study. To recap, it is important to highlight how the theoretical and conceptual background can strengthen the research study as follows:

- A clear understanding of the conceptual foundations gives the research structure and focus.
- Theories, concepts and assumptions drawn from existing knowledge can inform the formulation of hypotheses and the framing of the research question(s).
- They ground the researcher in the generalization of the broad facets of the phenomenon under study.
- This grounding of the theoretical and conceptual elements aids in determining the scope and boundaries of the study.

5.4   Literature Review and Analytical Framework

One of the key tasks for any researcher is to survey the scholarly literature relevant to the field of study. The researcher must interrogate current knowledge, discuss the relevant concepts and theories in greater depth and identify gaps in the existing literature on the topic. It is important to mention that unlike in the theoretical background, the literature review uses peer-reviewed journal articles and conference proceedings more extensively than books and grey literature. This distinction between the background discussion and the literature review is vital to good research design.

In most postgraduate research studies, researchers are encouraged to survey the scholarly articles and write reviews on the topic under investigation. This involves scan reading, critical reading and analysis, focusing on strengths, weaknesses, drawbacks, flaws and alternative points of view. In the sections that follow, we look at the literature review from two perspectives – the literature survey and critical literature reviews.

5.4.1   Literature Survey

The literature survey is the process of searching, finding and harvesting peer-reviewed journal articles and conference proceedings from reputable sources. The thrust of this phase of the research work is to situate the current study within the appropriate body of literature, identify leading articles and authors and ultimately provide context and direction to the research. A researcher requires a disciplined attitude and patience at this stage. To save time and to be productive, we recommend the approach in Figure 5.1, which shows a typical reading flow chart that can be handy since the researcher is faced with making decisions on which articles to read. The effectiveness of the researcher at this point is the ability to extract the most vital concepts, facts or ideas from the articles. Using the annotated bibliography technique discussed above will be of great assistance. In doing so, pertinent questions to ask yourself are:

- How relevant is the article to the topic? How do I rate the degree of relevance?
- How well does the article present and explain theories or concepts relevant to the topic? What can I learn from the explanation?
- To what extent is the author(s) writing and analysis objective and unbiased? How well does the article fulfil its own objectives? What value does it offer to my research?
- Is the article methodologically sound?

To address the above challenges, the researcher must read actively and deeply. Adhering to the flow in Figure 5.1, the first step is to skim the article with attention to the title, abstract and conclusions. This should be followed by in-depth study of the article, attempting to understand the author's argument, then making your own personal judgement of the effectiveness of the author's positions and noting aspects that are unclear or confusing. To recognize the gap(s) in knowledge requires good analytical thinking.

5.4.2   Critical Literature Reviews

A critical review is a directed appraisal of a selection of peer-reviewed journal articles, conference proceedings and scholarly papers, relating to the specific research question(s), accompanied by thorough comparative analysis of the ideas and themes derived from these related articles. Looking at the theories and concepts and how they are applied in the literature, the various shades of argument, the strengths and weaknesses in the articles, consider which articles and which ideas are most relevant to your research. The main focus is to infer how the ideas extracted from academic literature relate to your research question(s). It is important to think of how to review the articles in order to highlight key points, summarize the argumentation, propose new interpretations, emphasize perceptions and highlight conclusions relevant to your own research problem and research questions. A few of the key approaches to conducting a critical review are outlined below:

5.4.2.1   Theoretical Literature Review

The study should be underpinned by a particular theory, or theories, noting that there may have been an accumulation of relevant theories over time. The theoretical literature review attempts to examine the advancement of the relevant theories, drawing out the relevant concepts and assumptions about the phenomenon of study. So the researcher needs to establish what theories previously existed, the connections between them and the extent to which the theory has advanced. This is the foundation for generating new ideas in qualitative research or for hypotheses that require further testing in quantitative research. This focused review can reveal the lack of suitable theory or show that existing theory is inadequate for analysing new and evolving research questions.

5.4.2.2   Historical Literature Review

The idea of the historical literature review is to focus on examining the subject of study within a given timeframe or period. One example of a research component that would use the historical review approach would be understanding the evolution of cybersecurity measurement frameworks; see, for example, the following article:

Mbanaso, U., Abrahams, L., & Apene, O. (2019). Conceptual design of a cybersecurity resilience maturity measurement (CRMM) framework. *The African Journal of Information and Communication, 23*, 1–26. doi: https://doi. org/10.17159/2077-7213/2019/n23a2

The art of tracking the progression of a phenomenon within the context of a field of study can reveal gaps in the literature, creating the opportunity to search for new data, new analytical insights, new conclusions and possibly new theory. Placing research in a historical context can help to refocus the concepts to be applied differently in new research studies.

5.4.2.3   Argumentative Literature Review

In this type of literature review, the researcher evaluates the views expressed in selected literature in order to create opposition to a debate, a philosophical stance or a theoretical approach (Rowe, 2014). The aim is to develop a framework that attempts to establish an antagonist standpoint within the body of literature. This type of review typically requires the researcher to build argumentative capability and ways of scrutinizing the literature to form a unique perspective to guide the study.

5.4.2.4   Integrative Literature Review

The purpose of the integrative literature review is to build new frameworks based on the pragmatic integration of themes and concepts from a variety of sources, including systematic literature reviews written by other scholars, or critiques and syntheses of selected literature relevant to a study area. This entails examining related or similar hypotheses or research questions and drawing on the themes and concepts used in those studies. There are numerous sources of information for all literature reviews, including journal articles, conference proceedings, books, informal and formal expert reports, industry reports, government policy documents and other documents that are relevant to the field of study. Applying the funnel strategy framework discussed in Chap. 3 can help in limiting the number of sources used. It is vital to focus on the most recent sources and references (the last 3–5 years).

5.4.2.5   Methodological Literature Review

The methodological literature review is critical to the outcome and validity of a study. This kind of literature review is included in the methodology chapter, not in the literature review chapter. One of the steps research scholars take is to survey the methodological paradigms and approaches that have been applied to a particular topic or type of research problem, that is, a comparative survey of methodologies, commenting on the strengths and weaknesses of the methodologies applied. This enables the researcher to understand the methods and techniques used in various studies and draw inferences on what data collection and data analysis techniques to choose. In particular, the researcher needs to consider ontological and epistemological perspectives and the research methods for data collection and analysis (quantitative, qualitative or mixed methods), including sampling techniques, interviewing skills, survey design and other approaches. The methodological literature review uses (i) journal articles and books dedicated to the subject of research design and methodology and (ii) the same articles as used in the narrative or systematic literature review, but this time reviewing only the methodologies discussed in those articles.

5.4.2.6   Narrative Reviews Versus Systematic Literature Reviews

All the five types of literature review listed above can be conducted either as a narrative literature review or as a systematic literature review. Narrative literature reviews generally commence as random searches, with initial identification of search terms and with the researcher identifying additional articles from each new article read, until no significant new information is found. The literature is deconstructed to identify relevant themes, following which the narrative is developed and written up.

Systematic literature reviews follow a highly organized method. They are similar to but not the same as systematic reviews. Systematic reviews focus on extracting data from previous research on the subject, including published literature and unpublished research papers, and may include raw data or minimally processed data collected by other researchers. We will come back to the systematic review when we discuss data collection. On the other hand, the systematic literature review focuses almost exclusively on peer-reviewed journal articles and aims to intentionally document, summarize and critically examine the various factors, themes, concepts, models, frameworks and theories that relate to a well-stated research question. Below we set out a few of the initial steps to designing a systematic literature review (SLR):

- Main research question:

  Why is cybersecurity in public hospitals and clinics still at an early stage? = *Select search terms from the main question [cybersecurity in health, cybersecurity in hospitals, cybersecurity in primary healthcare]*

- Research sub-questions:

  Q1. How advanced are cybersecurity policies and frameworks in the case study environment?
  Q2. How advanced are cybersecurity technology and applications in this environment?
  Q3. How advanced are cybersecurity professional skills and user skills in this environment?

Research sub-questions {Q1–Q3} = *Select search terms for each of the research sub-questions. State the criteria for inclusion of articles in the SLR.*

The next step is to use these search terms to explore three or more journal databases, to identify a sufficiently extensive body of literature.

In the example below, on the subject of the use and impacts of e-health within community health facilities in developing countries, the initial search identified 265 journal articles. Following a review of the titles and abstracts, 184 articles were excluded as they did not fit with the criteria for inclusion. Then further 16 articles were included, after checking the lists of references in the 81 articles initially included, meaning that 97 full-text articles were included in the detailed review (reading) process. During the review process, 84 articles were excluded for the reasons stated in the diagram; thus, only 13 articles were included in the analysis of the literature, as illustrated in Figure 5.5.

Students should read extensively on the systematic literature review tools and techniques, for example:

Kitchenham, B., Brereton, O., Budgen D., Turner, M., Bailey, J., & Linkman, S. (2009). Systematic literature reviews in software engineering – A systematic literature review. *Information and Software Technology, 51*, 7–15. doi: https://doi.org/10.1016/j.infsof.2008.09.009

Mahmud, K., & Usman, M. (2018). Trust establishment and estimation in cloud services: A systematic literature review. *Journal of Network and Systems Management, 27*, 489–540. https://doi.org/10.1007/s10922-018-9475-y

Merli, R., Preziosi, M., & Acampora, A. (2018). How do scholars approach the circular economy? A systematic literature review. *Journal of Cleaner Production, 178*, 703–722. https://doi.org/10.1016/j.jclepro.2017.12.112

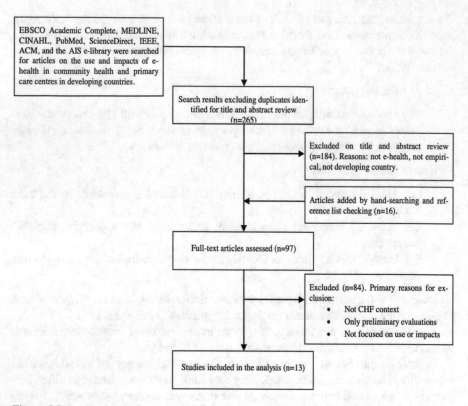

Figure 5.5 *Systematic Literature Review Process*. Source. Cohen, Coleman and Abrahams (2015), page 4

5.5   Structuring and Organizing the Literature Review

The important ways of creating a well-constructed literature review include the following:

5.5.1   Locating a Focal Point or Central Interest

The literature review must relate directly to your research question, which arises from the research problem statement. The literature review should present between three (for Masters research) and four/five themes (for PhD research), related to each of the research sub-questions. The main research question and the sub-questions provide a focal point or central interest. Choose the type(s) of literature review you are writing. Section 5.3.1 has outlined review types, which can guide the researcher to focus on the perspectives to be understood, instead of simply listing all the sources and writing details about each one of them sequentially.

It is important to read broadly and thoughtfully select current articles that meet the desired review criteria. Select articles that address similar concepts, theories and assumptions, so that the points of argument, approaches, methods and philosophies can be comparatively analysed and presented. Consider the following: How well does the author(s) present the data and analysis? How well is the study conceptualized and is their theoretical clarity? Does the article reveal important thinking in the field? Are the lines of argument thoughtful and open to question?

When writing a literature review, it must be structured to convey the ideas clearly, concisely and logically to the audience. For example, when writing a literature review on web congestion for the Internet of Things (IoT), the researcher can use the mind mapping approach, described in Chap. 3, to focus on writing an integrative literature review.

5.5.2   Logical Coherence

A coherent literature review will give clarity to the research problem, the main research question and the research sub-questions. The researcher should decide the most effective way of presenting the review, with attention to the following guiding points: What are the most important concepts that the review needs to include? In what order should they be presented? The concepts should always be discussed in the same logical sequence.

5.5.3   Basic Structural Components

Similar to conventional academic papers, writing literature reviews follows an organized structure. It comprises a short introduction section (overview of the literature review chapter) and the body of the review containing the discussions of the themes of the review (conceptual and theoretical perspectives in the case of qualitative studies, as well as assumptions in the case of quantitative and mixed methods studies), followed by a chapter summary or chapter highlights. Depending on the type of review, strengths, research gaps, weaknesses, flaws, etc. should be included in each thematic section of the body of the chapter. A valuable inclusion is the visualization of the analytical framework in diagram format. In the chapter summary section, the researcher should highlight the direction of thinking of the present study in contrast to the articles reviewed.

The following provides a brief description of the content and style of the literature review:

Introduction: This section gives a quick overview of the subject matter under review, highlighting the research area; the central theme of the review, for example, web congestion control; and the literature review structure.

Body of literature: This section deals with detailed discussions on the identified theories and concepts, arranged chronologically, thematically or methodologically (as explained below). Organizing the body of the literature review chapter is necessary to identifying the compelling research gaps. Use interesting sub-headings to sign-post the key points in the literature review. Once the components and sub-headings are settled, presenting the ideas derived from the reviews becomes the next stage of concern. To organize the structure, the writing should always follow a logical sequence, which can be one of the following approaches:

Chronological order/history: The researcher presents the ideas according to when they were prominent in the literature, showing a periodization, for example:

> Period 1: Early-stage web development, no congestion control (year x to year y)
> Period 2: Shift to the informational web, minimal congestion control (year xx to year yy)
> Period 3: Shift to the transactional web and e-commerce, high congestion control (year xxx to year yyy)
> Period 4: Shift to the Internet of Everything, maximum congestion control (year xxxx to year yyyy)

Examining global and local trends in theory: Another effective way of organizing the review is to examine the theoretical trends in the literature, for example:

- Theoretical insights relevant to fintech innovation and adoption
- Theoretical insights relevant to fintech innovation and adoption on the African continent
- Theoretical insights relevant to fintech innovation and adoption in Country X

Thematic differentiation: In this approach, literature reviews are organized and presented based on concepts or themes, rather than the progression of time or trends. The progression of time may still be important to include in a thematic review, but the thematic review should dominate (Theme A, Theme B, Theme C), noting that a strong thematic review tends to break away from chronological order (Dewey & Drahota, 2016).

Methodological order: This review approach focuses only on the methods used by the researcher and not on the content of the material.

5.6   Analytical Framework Diagram

The literature review should end with an analytical framework diagram, in which the periods, trends, themes or methodologies are visualized and synthesized. The diagram in Figure 5.6 demonstrates the comprehensive cybersecurity vulnerability landscape, guiding the researcher with respect to the areas for data collection and laying the foundation for data analysis.

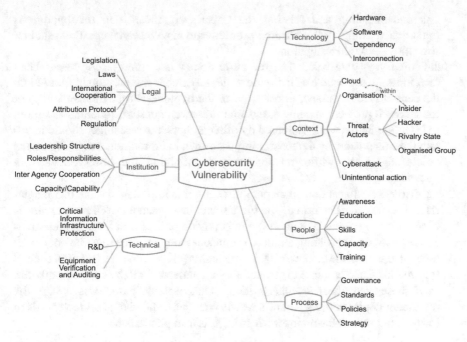

Figure 5.6 *Cybersecurity Vulnerability Conceptual Landscape*

5.7   Important Hints on Writing Literature Reviews

There are a few guidelines to be used when writing the review, as follows:

Evidence-based arguments: The researcher should always ensure that the line of
argument, or point made, is supported by a reference(s). What will substantiate
the argument, or point, or claim, or interpretation, is how the researcher justifies
the positions she/he takes with the necessary evidence to demonstrate a valid
argument.

Selectiveness: A researcher should be familiar with the work of the leading scholars
in the subject area of research, always check the views of the authors and select
the most important ideas in each source to highlight in the review. The choice of
information should be relevant and directly related to the review's focus, whether
it is chronological, thematic or methodological.

Quote sparingly in the review: Avoid the use of quotes for referencing unless it is
unavoidable, but even in this case, use reasonably short quotes. Always use quo-
tation marks and the in-text reference (citation), including page numbers for
direct quotes. You are being assessed for your capacity to present ideas in your
own words, not to quote other authors. The rule of thumb is 'cite, don't quote'.

Summary and synthesis of each article: The researcher is expected to present the
central features of the article/paper under review, *in your own words*. It is important to synthesize the contribution or value of the article by showing the research-

er's understanding and relating the paper's significance to the hypothesis (quantitative study) or the research problem and main research question (qualitative study) under investigation.

Maintaining your own voice: The literature review is centred on the researcher's own ideas. In the course of the review, the researcher should remain focused on the central idea discussed, as reflected in the hypothesis (quantitative study) or the research problem statement and main research question (qualitative study). In highlighting the work presented by other authors, the researcher should maintain their own voice by demonstrating their own interpretation and views on the article highlights, while constructively highlighting the observed gaps in knowledge.

Using discretion and caution when paraphrasing: It is important to note that plagiarism is a very serious crime, showing fraudulent behaviour. The researcher is expected to represent the author(s)'s information or perspectives in the researcher's own words. Presenting an idea or concept from a source must be done with thoughtfulness while acknowledging the author(s) by proper in-text referencing. It is unethical and criminal to present another author's work by copying verbatim as if these were the researcher's ideas. Unacceptable paraphrase, where the researcher copies another author's words and ideas, and changes a few words in the text, but retains the major word strings, is also plagiarism.

5.8   Chapter Summary

This chapter has explained the foundational research skills for research design and academic writing, as well as the tools and techniques for writing the background discussion and the literature review. This aims to equip the researcher with the capacity and skills needed at the early stage of scholarly study. The skills component presented the soft skills that can accelerate effectiveness and productivity for a researcher. The background discussion is necessary for the researcher to understand the real-world environment and the theoretical foundations that underpin the study. The literature review helps the researcher to understand the concepts and theories in greater depth and their particular relevance to the research problem/questions and shapes the direction of the new study.

5.9   Chapter Exercise

This exercise requires you to write an annotated bibliography for only three articles, but in practice, the researcher will write an annotated bibliography for between 30 and 80 articles, depending on whether the literature review is for a coursework Masters, a research Masters or a PhD.

Instructions:

1. Write the first annotated reference for the first article, then move on to the second article, finally the third article.
2. First part of the annotation: Write the correct APA style reference (as if this were in the list of references) (use the free online APA style tutorial and APA style blog available at www.apastyle.org).
3. Second part of the same annotation: Write the highlights of key points from the article (150–200 words).
4. Third part of the same annotation: Write the analytical commentary on the highlights of key points from the article (150–200 words).
5. After completing all three annotations: Write the literature review statement about the three articles in relation to each other, building on the three analytical commentaries you wrote (600 words).

Bibliography

Abend, G. (2008). The meaning of 'theory'. *Sociological Theory, 26*(2), 173–199. http://www.jstor.org/stable/20453103

Adom, D., Hussein, E., & Agyem, J. (2018). Theoretical and conceptual framework: Mandatory ingredients of a quality research. *International Journal of Scientific Research, 7*(1), 438–441. https://www.researchgate.net/publication/322204158_THEORETICAL_AND_CONCEPTUAL_FRAMEWORK_MANDATORY_INGREDIENTS_OF_A_QUALITY_RESEARCH

Babbie, E. (2016). *The practice of social research* (14th ed.). Cengage Learning. [Available as an e-book].

Cohen, J., Coleman, E., & Abrahams, L. (2015). Use and impacts of e-health within community health facilities in developing countries: A systematic literature review. *ECIS 2015 completed research papers* (Paper 33). http://aisel.aisnet.org/ecis2015_cr/33

Cryer, P. (2006). *The research student's guide to success* (3rd ed.). Open University Press.

Dewey, A., & Drahota, A. (2016). Introduction to systematic reviews: Online learning module. *Cochrane Training*. https://training.cochrane.org/interactivelearning/module-1-introduction-conducting-systematic-reviews

Imenda, S. (2014). Is there a conceptual difference between conceptual and theoretical frameworks? *Journal of Social Science, 38*(2), 185–195. https://doi.org/10.1080/09718923.2015.11893249

Kitchenham, B., Brereton, O., Budgen, D., Turner, M., Bailey, J., & Linkman, S. (2009). Systematic literature reviews in software engineering – A systematic literature review. *Information and Software Technology, 51*, 7–15. https://doi.org/10.1016/j.infsof.2008.09.009

Mahmud, K., & Usman, M. (2018). Trust establishment and estimation in cloud services: A systematic literature review. *Journal of Network and Systems Management, 27*, 489–540. https://doi.org/10.1007/s10922-018-9475-y

Mbanaso, U., Abrahams, L., & Apene, O. (2019). Conceptual design of a cybersecurity resilience maturity measurement (CRMM) framework. *The African Journal of Information and Communication, 23*, 1–26. https://doi.org/10.17159/2077-7213/2019/n23a2

Mensah, S., et al. (2019). *2021 banking sector outlook: Nigerian banks face struggles on many fronts*. Standard & Poor's Financial Services LLC. https://www.spglobal.com/_assets/documents/ratings/research/100049447.pdf

Merli, R., Preziosi, M., & Acampora, A. (2018). How do scholars approach the circular economy? A systematic literature review. *Journal of Cleaner Production, 178*, 703–722. https://doi.org/10.1016/j.jclepro.2017.12.112

Rowe, F. (2014). What literature review is not: Diversity, boundaries and recommendations. *European Journal of Information Systems, 23*, 241–255. https://doi.org/10.1057/ejis.2015.7

SABRIC. (2019). *SABRIC annual crime stats 2019*. https://www.sabric.co.za/media/1265/sabric-annual-crime-stats-2019.pdf

Serianu. (2017). *Nigeria cybersecurity report 2017: Demystifying Africa's cybersecurity poverty line*. https://www.serianu.com/annual-reports.html

Serianu. (2018). *Africa cybersecurity report – Kenya: Cybersecurity skills gap*. https://www.serianu.com/annual-reports.html

Serianu. (2019). *SACCO cybersecurity report 2019: Digital transformation and cyber risk within SACCOs*. https://www.serianu.com/industry-reports.html

Serianu. (2020a). *Africa cybersecurity report Kenya, 2019/2020: Local perspective on data protection and privacy laws: Insights from African SMEs*. https://www.serianu.com/annual-reports.html

Serianu. (2020b). *Africa cybersecurity report Uganda 2019/2020: Local perspective on data protection and privacy laws: Insights from African SMEs*. https://www.serianu.com/annual-reports.html

Swanson, R. (2013). *Theory building in applied disciplines*. Berrett-Koehler Publishers.

Chapter 6
Research Philosophy, Design and Methodology

6.1 Introduction

Research philosophy, design and methodology are critical components of scholarly research; hence, it is important to examine each of these components carefully in the preparation of computer science (CS), information systems (IS) and cybersecurity (CY) research studies. We need to understand where and how each of these elements fits into the research process. Ideas about research philosophy, research design and research methodology relate to the conceptual thinking about what constitutes new knowledge and how to produce new knowledge, as well as the ethics applied when engaging in research. A researcher's philosophical choices shape the assumptions, perceptions and interpretations of the nature of reality. Our individual research philosophy in relation to a particular research project influences how we see the research problem, how we understand data, what analysis we make, what conclusions we draw and what theories we advance. For example, in some studies, we may take a human-centric approach, in other studies, we may take a techno-centric approach and on occasion we may balance the two in a techno-human approach. Consequently, a researcher must be grounded in understanding the need to adopt a clearly stated research philosophy, early in the research process. Moreover, these three elements form the structural basis that examiners use to evaluate and validate a research study. Typically, an inappropriate research philosophy, design and methodology may result in invalid study outcomes, where the interpretations and conclusions are faulty. In this chapter, the rudiments of research philosophy, design and methodology are explained within the domains of CS, IS and CY. Babbie (2010) is a good companion text to this chapter.

6.2   Research Philosophy

Research philosophy is primarily concerned with cognitive theory and its relevance to creating and expanding knowledge (Novikov & Novikov, 2012), in other words, creating and expanding what we know about any aspect of the universe. Our knowledge of reality can be perceived from mainly four paradigms – positivism, interpretivism, pragmatism and realism (Saunders et al., 2019). A researcher must aim to understand the philosophical dimensions of the research issues, in order to select the research paradigm in which to think, which is required to understand and solve the problem (Mkansi & Acheampong, 2012). The researcher must consider how a research problem should be viewed, understood, conceptualized and resolved. Figure 6.1 depicts the rudiments of research philosophy. The researcher's philosophical assumptions start from the ontological assumptions, then move to epistemological assumptions, then onto axiological assumptions, doxological assumptions and finally to methodological assumptions. All of these philosophical choices influence the research design and therefore the research findings.

Each of these types of philosophical assumptions is further expounded upon below:

6.2.1   Ontology

Ontology implies an understanding of, or a specific set of insights into, the realities of the observable world. In this sense, reality is an interpretation of the researcher's understanding of the problem and of the researcher's interpretation of the data

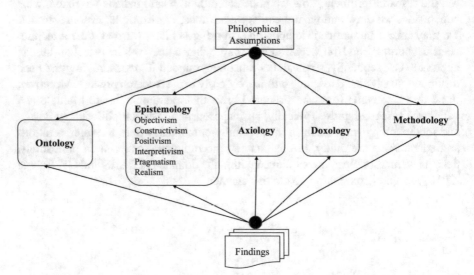

Figure 6.1   *Rudiments of Research Philosophy*

collected. In CS, IS and CY, ontology relates to the axioms, characteristics, properties, representations, categories and propositions that underlie how reasoning is performed in these domains. In simple terms, it is a shared conceptualization of underlying concepts of the subject domains and the relationships that exist between them. This means that ontology is concerned with theorizing the reality of a particular phenomenon, providing a 'systematic description' of the ideas and terminology used to understand the research problem (Kaspersky Daily, 2021). Consequently, it is important that the researcher builds a small repository of the relevant ontologies applicable to the topic of study, as the basis for generating a deep understanding of the topic (Saunders et al., 2019).

6.2.2 Epistemology

Epistemology relates to the intellectual tools we use to confirm that the research produced constitutes knowledge. It is necessary to provide evidence that the nature of the research design leads to knowledge, not to conjecture. It is necessary to provide evidence that what the researcher claims is knowledge is indeed knowledge. In their explanation of the 'theory of knowledge' underpinning a research study, Sallos et al. (2019, p. 583) argue that 'an explicit, context specific epistemological foundation is essential for the reconciliation of the variety of perspectives and dimensions of cybersecurity management and strategy'. What this means is that the researcher has to make certain decisions and certain choices, when preparing the research design. For example, if the research aims to understand the nature of cybersecurity professional skills in an organization (phenomenon A), or in the banking sector (phenomenon B), then the research design should adopt an explicit, qualitative, constructivist, case study approach, which will enable the researcher to collect data that provides deep insight into the particular context. On the other hand, if the research aims to measure the extent of cybersecurity awareness among customers of a group of banks, then the research design should adopt an explicit quantitative survey approach, which will enable the researcher to measure various levels of cybersecurity awareness and to deduce what proportion of banking customers have a low, medium or high level of cybersecurity awareness. From this brief explanation, we can see that the scope of the research problem and the application of problem-specific methods can increase the validity of the results. Simply, epistemology is concerned with justification for the claim that the research submitted for examination, or for publication, constitutes knowledge. In the fields of CS, IS and CY research, it is important to state the epistemological grounds for your study, whether the study applies formal methods of logic, or probability theory, or computability theory or other theoretical foundations.

How the researcher approaches, designs, collects and analyses data and then concludes the study can be based on positivism, interpretivism, pragmatism or realism; see Figure 6.2.

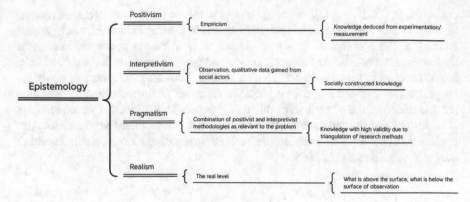

Figure 6.2 *Epistemology*

6.2.2.1   Positivism

Positivism relates to the rationality of the views expressed, demonstrating that the knowledge generated in the research process exists independently of what is in the mind of the researcher. For CS, IS or CY research studies, the phenomenon may be studied empirically, by experimentation or by observation of the problem in its setting but must then be objectified through using scientific concepts or theories formulated by the researcher, to express the nature of the phenomenon. Empiricism does not rely on existing theory but rather on the data the researcher acquires during the experimental process. Therefore, the scientific design of the experiment, or the observation, is important as it influences which data will be included in the study.

6.2.2.2   Interpretivism

Alternatively, interpretivism incorporates a range of methodologies, tools and techniques aimed at understanding social reality, including how organizations function. Researchers can interpret the social nature of the phenomenon by reviewing relevant documents (such as public policy documents, minutes of meetings/events, internal rules of the institution and others), as well as by conducting interviews to collect data from people who have a close acquaintance with the phenomenon of study. Constructivism is an important form of interpretivist research where the data is gathered from human subjects. These are interview respondents who have knowledge or experience of the research problem, and the analysis and conclusions are strongly shaped by the respondent's reports of the lived experience of the phenomenon. According to Elkind (2005, p. 334), 'constructivism is the recognition that reality is a product of human intelligence interacting with experience in the real world. As soon as you include human mental activity in the process of knowing reality, you have accepted constructivism'.

6.2.2.3   Pragmatism

Pragmatism is a problem-oriented approach to research, where the investigator chooses whatever research methodologies may be required to understand the problem, often a combination of methodologies. Thus, the researcher may use mixed methods, including quantitative and qualitative approaches, more than one quantitative method or more than one qualitative method. The researcher may, for example, first employ a quantitative survey to measure the extent of the problem and then use qualitative interviews to investigate a specific aspect of survey data more deeply. Hence, pragmatism combines the most useful aspects of positivism and interpretivism, as may be necessary.

6.2.2.4   Realism

Realism, briefly, is an approach to research that considers it necessary to get as close as possible to understanding the nature of what is real in relation to the research problem. The realist is not satisfied with the limited data that can be gained from positivist and interpretivist studies, and the realist is not satisfied with getting as close as possible to understanding the problem by using a pragmatic approach. The realist wishes to examine the problem in some depth and therefore is most likely to select a constructivist case study methodology, as a means of gaining deep insight into the problem. The realist understands that a quantitative survey is unlikely to reveal the layers of the problem and that a qualitative survey of many instances of the problem is likely to only reveal a superficial understanding of the problem, in other words limited data across a large sample. The realist seeks to study a smaller sample, or sample population, in depth, to reveal as many facets of the problem as may be possible. Such studies are particularly useful and appropriate where the research problem is in an institutional setting, for example understanding the strengths and weaknesses of traditional enterprise resource planning (ERP) systems in an institutional context when the market is moving to next-generation ERP systems; or understanding why professional cybersecurity skills development is weak in the financial services sector.

In this section, we have only examined positivism, interpretivism, pragmatism and realism, as these are a good foundation for the novice researcher. However, it should be understood that there are many types of epistemology; thus, the student should explore the field of research philosophy as widely as possible in order to select the appropriate research philosophy.

6.2.3   Axiology

Axiology refers to the nature of the value produced in a research study, guiding the researcher to discuss and explain whether the value or significance of the study is of an economic nature, of a social nature and of an institutional nature, relevant to a particular narrow audience, to a broader audience. In CS, IS and CY research, axiology is a necessary component for evaluation of the validity of the study. In computer science, axiology may simply refer to the value of particular algorithms or computing code or the value of additions to the body of knowledge in computational mathematics. In information science, read the short article by Gladden (2017), who discusses an axiology of information security for human health devices. In cybersecurity, axiology could explain the value of specific research into cybersecurity policy and regulation.

6.2.4   Doxology

Doxology refers to understanding what is required to formulate a 'believable' research outcome. Why would examiners or readers believe the analysis, the conclusions, the theories generated in your research? Goldman (2001) discusses the relationship between doxology (theories of how beliefs form) and epistemology (theories of how knowledge forms). The reason why it is important for a researcher to think about and write about how beliefs form is that the researcher needs to understand why a reader should believe the evidence, analysis and conclusions of the research produced, in other words, what is it about the research design and outcomes that makes it possible to believe in the results. This is a way of checking that the researcher has met some of the requirements for examination. Simply believing is not satisfactory at all. If the examiner is a cybersecurity expert, they will only believe in the outcomes if the work is epistemologically defensible and if the work is interesting and engaging in relation to their particular expertise and challenges them to reconsider previously held beliefs. If the reader is an industry practitioner, the reader may only find the research results believable if they can relate the results to their own industry experience. In contemporary quantitative research, doxology is expressed in the validity of the theoretical assumptions and in the formulation of the hypothesis, which the research needs to test to verify and validate it as 'knowledge'.

6.2.5   Methodology

Methodology is about the nature of the research process, in other words how the researcher applies particular methods, principles and techniques to identify, select, collect, process and analyse data and draw conclusions in a field of study (Sileyew, 2019, Chap. 3). Choosing a methodology requires the researcher to consider whether a quantitative, qualitative or mixed-methods methodology is suitable to collecting and analysing data, given the nature of the research problem.

6.2.6   Why Is Philosophical Reasoning So Important in Research?

This is because ontology and epistemology create a holistic view of how knowledge is derived, developed and viewed and how researchers think about the knowledge they are creating, providing reasons why we should trust the quality of the research. Philosophical assumptions will enhance the quality of research and can contribute to the creative thinking of the researcher and to the ingenuity and validity of the research. Research philosophy enables clarity in the research design, guiding the researcher to identify the suitable design that should work and avoid designs that are not feasible. Research philosophy is continuously advancing, particularly in relatively new fields such as cybersecurity, where the reader needs to be given the basis for critically evaluating a study's overall validity and reliability. In the next part of this chapter, we offer ideas to deepen the understanding of philosophical paradigms, ideas which the researcher should relate to the particular CS, IS and CY study being conducted.

6.2.6.1   Perspectives on Epistemology: Positivism

Positivism is used to explain a scientific approach that is based on experiments and statistics to study the natural world or to study particular aspects of society based on social experiments or measurement. Positivism can be seen from the perspective of determinism, empiricism and generalisability (Cohen et al., 2011). The notion of 'determinism' refers to events, or effects, that result from specific circumstances, such as cyber-attacks that result from flaws and vulnerabilities in an information ecosystem. The understanding of such casual links is indispensable for forecasting what is likely to occur in the future if the circumstances remain unchanged. In the context of positivism, 'empiricism' describes the collection of provable experimental data that can support hypotheses or theories generated by the researcher. A good example is the use of surveys to support the understanding of information system users' behaviour. The positivist can use the concept of 'generalisability', which refers to designing a study and choosing a methodology that enables the researcher to generalize the findings for a particular narrowly defined study to a broader context of similar research problems. For example, the results of a cybersecurity case study of a specific commercial bank could be extended to all commercial banks, on condition that the research design and methodology enables such generalisability.

   Positivism depends on quantifiable observations that can lead to statistical analysis. Positivism has 'an atomistic, ontological view of the world as comprising discrete, observable elements and events that interact in an observable, determined and regular manner' (Collins, 2010, p. 38). Consequently, most positivist research approach adopts quantitative methodology and deductive reasoning, inferring that positivism depends on 'things' that are quantifiable through observations or experiments, and uses statistical analyses to draw conclusions; see Table 6.1. While

Table 6.1 *The Positivist Paradigm*

	Ontology	Epistemology	Theoretical perspective	Methodology	Methods
Paradigm	The real nature/ characteristics and existence of the CS, IS or CY phenomenon	An understanding of why specific data and analysis about the phenomenon constitute valid knowledge	Which theoretical perspectives apply to the research problem	A systematic approach that enables the researcher to apply a single method or a combination of methods	The specific techniques used to collect and analyse data
Positivism	There exist a number of possible characteristics about the phenomenon, where only specific characteristics become apparent as a result of the particular positivist research design	Quantitative data and statistical analysis often provide mainly high-level descriptive data, though some analytical techniques lay the basis for predictive analysis	Theoretical perspectives are founded in statistical analysis and statistical models or in experimental thinking	Experimental, correlational, longitudinal and other forms of modelling or survey-based research	Quantitative sampling, measurement-based data collection and data analysis techniques

positivist paradigms can apply to CS, IS and CY, most studies in computer science will adopt a positivist philosophy.

Research conducted in the positivist paradigm can enable a level of predictiveness in relation to future events, based on experimental evidential data that tests a hypothesis to either refute or prove ideas about the phenomenon. In other words, scientific quantification can be the basis for predicting cybersecurity occurrences, based on specific cause-and-effect parameters. The researcher can improve the accuracy of prediction by improving the parameters based on previous studies of the same phenomenon.

6.2.6.2 Perspectives on Epistemology: Interpretivism

Interpretivism, often viewed as the opposite of positivism, argues that the social fabric of human enterprise is overarching and complex (Mkansi & Acheampong, 2012) and can best be understood by understanding the many layers of complexity. While positivism lends itself to well-defined 'laws' or 'models' or 'formulae', interpretivism claims that critical understanding of our complex world is not possible if such complexity is reduced entirely to a set of measurements (Cohen et al., 2011). Interpretivism requires the researcher to understand how the 'social actors', or 'human subjects', see the phenomenon of study. Interpretivism can be viewed from multiple perspectives in social science inquiry, from which we examine the three most common perspectives, namely, phenomenology, constructivism and symbolic interactionism.

Phenomenology argues that research insight is a function of the experience gained by the researcher's direct interaction with the particular events or phenomenon. It argues that the researcher cannot access the objective external reality directly and can only interpret the phenomenon based on observation. Rasmussen (1998, pp. 554–557) explains that phenomenology is about how the researcher perceives the object of study, not about how the object actually is. The researcher, as observer, influences what is reported about the object of study, because the researcher has a particular consciousness which may perceive the object in a particular way. However, the researcher can also intentionally shift their view of the object, to better understand the phenomenon or to perceive more dimensions of the phenomenon.

Constructivism expresses the idea that researchers cannot experience or observe the reality of the phenomenon directly and therefore need to engage with key informants who can share their experience of the phenomenon and with documents that provide data about the phenomenon. For example, if the researcher wishes to understand the application of cybersecurity laws by the courts, then the researcher will need to interview legal practitioners, judges and court officials, who will share their lived experience of the law or of the strengths and weaknesses of the legal environment. Phenomenology and constructivism are both forms of interpretivism, but constructivism relies on documentary evidence and on the perceptions and perspectives of key informants, rather than on the observations of the researcher (Rasmussen, 1998, pp. 557–559). Thus, constructivist research enables us to gain some insight into the phenomenon of study but does not give us an exact replica of the reality of the phenomenon. This means that the researchers must provide clear analytical argument for the interpretation of the data collected.

Symbolic interactionism emphasizes that people interpret their experiences in terms of a shared understanding of social phenomena, communicated through language or other media, where meaning is co-created by social groups or by institutional groups (Harvey, 2012–2020). Thus, when the researcher interviews key informants, the difficulty in interpreting the phenomenon of study is that the researcher must decode the symbolic meaning that the informants have expressed. In cybersecurity research, for example, individuals participating as respondents in case study research may regard password protection as low risk and therefore low priority for an organization, whereas the researcher may understand that password protection is medium to high risk in that sector or in another sector, and the researcher must therefore interpret whether the key informants have good reason for their views and must explain why the respondents hold views contrary to conventional wisdom.

In concluding this section, Table 6.2 summarizes the interpretivist paradigm.

In general, the interpretivist paradigm will apply more to IS and CY than to CS inquiry, since the generation of new knowledge in CS requires experimental evidence or measurable data to validate what is claimed about the phenomenon. On the

Table 6.2 *Interpretivist Paradigm*

	Ontology	Epistemology	Theoretical perspective	Methodology	Methods
Paradigm	The real nature/ characteristics and existence of the CS, IS or CY phenomenon	An understanding of why specific data and analysis about the phenomenon constitute valid knowledge	Which theoretical perspectives apply to the research problem	A systematic approach that enables the researcher to apply a single method or a combination of methods	The specific techniques used to collect and analyse data
Interpretivism	Reality is open to interpretation as people express their own experiences and perspectives of reality	Qualitative data and analysis provide data that is open to interpretation requiring thick description and argumentation to explain data analysis	Theoretical perspectives are founded in qualitative theory building	Approaches that enable in-depth investigation and multidisciplinary or interdisciplinary research	Interviews, focus groups, document review and many other qualitative data collection and data analysis techniques

other hand, information systems research and cybersecurity research study, phenomena in ways that are less precise and are open to interpretation.

6.2.6.3 Perspectives on Epistemology: Pragmatism

Pragmatism is philosophical thinking that holds the perception that practicality and workability are important criteria for selecting the approaches and techniques with which to investigate a phenomenon (Mkansi & Acheampong, 2012). It is possible that a research question may not be addressed by a single research philosophy; it may take more than a philosophical perspective to understand and answer a particular research question. Pragmatism suggests that when it is unclear that either a positivist or an interpretivist philosophy alone can help to answer the question, it is feasible to adopt a mixed-methods approach, combining both qualitative and quantitative paradigms as appropriately as possible to answer the research question (Morgan, 2013). Table 6.3 summarizes the pragmatist view. The pragmatist approach can apply to IS and CY studies in relation to research that relates to technology, process and people. For instance, in creation and design research, producing an artefact can span more than one philosophical approach, hence the researcher may use qualitative and quantitative approaches in different phases of the research.

Table 6.3 *Pragmatism*

	Ontology	Epistemology	Theoretical perspective	Methodology	Methods
Paradigm	The real nature/ characteristics and existence of the CS, IS or CY phenomenon	An understanding of why specific data and analysis about the phenomenon constitute valid knowledge	Which theoretical perspectives apply to the research problem	A systematic approach that enables the researcher to apply a single method or a combination of methods	The specific techniques used to collect and analyse data
Pragmatism	Understanding that both quantitative and qualitative insights is needed to express the nature of the phenomenon	Understanding that quantitative and qualitative techniques when combined enhances the validity of knowledge	Drawing on theoretical insights from the quantitative and qualitative domains to enrich the foundational theory for the particular study	Applying mixed methods in ways that complement each other	Using a combination of data collection and analysis methods, for example, surveys and interviews

6.2.6.4   Perspectives on Epistemology: Realism

Realism asserts that things which exist, whether in nature or in social settings, are independent of whether it is recorded or perceived that they exist (Schwandt, 2015). Realism is built on sets of assumptions about what is required to explain 'real' causes, structures, processes or objects, on the understanding that these assumptions enable researchers to explain the physical world, as well as the social world. Realism assumes that scientifically, theories and hypotheses about the underlying causes of observable phenomenon can be appropriately formulated to derive explanations of what is observed (Lakoff, 1987). While some phenomena may not be observable at a particular point in time, they may become observable as new methodologies or tools become available. An example is that computer viruses were once unobservable but are now observable using various forms of anti-virus analytical tools based on software research and development (R&D). Consequently, it can be said that realist research philosophy depends on the notion of the 'eccentricity of reality' that is independent of human thought (Haig & Evers, 2015), meaning that there is no specific perfectly observable, perfectly describable reality, but researchers can observe particular aspects of the real world, in particular contexts, using particular methodologies and instruments. If, in research philosophy, objects are ontologically independent of the researcher's conceptual capacity, insights, opinions and beliefs, then the realist research must take great care in describing and explaining what is regarded as real. The words, terminologies and concepts applied in realist research must be carefully chosen to reflect as closely as possible what is being observed, rather than simply expressing the views of the researcher. This is important because two or more researchers will observe two or more 'realities' when observing the same phenomenon, hence the need to aim for some level of precision in description and explanation. If care is taken in the use of language of description and explanation, then many researchers, observers and practitioners can benefit from the knowledge created, partly because there is the possibility of creating a common understanding of the phenomenon under investigation.

Realism can be categorized into two thematic areas, namely, direct realism and critical realism. Direct realism, also referred to as naive realism, is best described as what you see is what you get (Schwandt, 2015), which implies that direct realism portrays phenomena through personalization of what is observed. Critical realism, on the other hand, claims that the real world is perceived by humans through experience due to sensations and images of real objects or processes. However, it argues that sensations and images of real objects or processes can be illusory and may not necessarily describe specific details of the real world. An example is where the researcher is observing a cybersecurity training session. The observation may suggest that the participants are highly knowledgeable, but this may be because they attended foundational training in the past week and have only just acquired the knowledge, not that they have acquired this knowledge over an extended period and are effective cybersecurity professionals. The direct realist would deduce that the participants are highly knowledgeable, while the critical realist will seek a more detailed explanation for why the participants express their cybersecurity knowledge so well.

6.2.6.5 Quick Comparative Overview of Epistemology

Understanding the principles of research philosophy can guide most efforts at research design. Thus, adopting a research philosophy is a critical step for planning and conducting research. While positivism stands for objectivity, measurability, predictability, controllability, laws of nature and rules of human or physical behaviour, non-positivist research philosophies stress understanding and interpretation of phenomena to make meaning out of the events. Table 6.4 presents selected paradigms and how they may influence studies in CS, IS and CY.

To summarize, research philosophy is concerned with how the researcher views the research problem under investigation and what ontologies, epistemological thinking and axiological thinking the researcher contemplates and uses in the research design. It provides the underlying explanation of the nature of knowledge and considers how to understand phenomena leading to the discovery of new knowledge. It provides the basis for explaining that the way in which the particular research study is conducted leads to a reasonable degree of trust in the knowledge produced.

Table 6.4 *Applying Epistemology in CS, IS and CY*

	Computer science (CS) research	Information systems (IS) research	Cybersecurity (CY) research
Positivism	Positivist approaches dominate research in this field with significant use of hypothesis testing and experimentation using scientific and empirical methods. The post-positivist acknowledges that quantitative research is influenced and informed by the researchers' own ideas. The research design is usually quantitative, applying deductive reasoning	Positivist approaches influence studies in cases where there is a need to determine 'cause and effect', especially in predictive models where quantitative dependent and independent variables are central to the studies	Similar to IS research, positivist approaches inform the design of studies where the goal is to control 'cause and effect', particularly in analytical models that influence how quantitative dependent and independent parameters affect the outcome of the studies. Positivist approaches are also suitable for the design of quantitative frameworks to measure cybersecurity awareness levels or maturity levels
In most descriptive research design, the positivist is influenced in the construction of the research question by phrases such as the following: How many? How frequently? What percentage? What proportion? To what extent? The outcome is usually a set of numeric values that can be statistically analysed to draw inferences and comparisons. Quantitative research questions commonly set the scene for the whole study. Three main types of research questions guide quantitative research designs: descriptive, comparative and relationship-based including 'cause-and-effect' research questions			

(continued)

Table 6.4 (continued)

	Computer science (CS) research	Information systems (IS) research	Cybersecurity (CY) research
Interpretivism	Interpretivist approaches rarely influence studies in this field unless they are building on initial quantitative research or experimental design	Interpretivist approaches can influence studies in the IS field, especially where researchers seek understanding of the world based on subjective experiences, often using a constructivist paradigm. Particularly in evaluating IS users' behaviour, researchers can enquire about the complexity of such behaviour. In this context, semi-structured, open-ended interview questions invite respondents to frame their understanding of the events, situations and phenomena	Interpretivist approaches are strongly influential in cybersecurity research, where constructivist case studies enable researchers to focus in depth on the processes and interactions that shape organizational responses to cybersecurity vulnerabilities. A typical example: in evaluating why users easily fall prey to commercial engineering attacks, the researcher can inductively develop a pattern of meaning, leading to theory building

The interpretivist usually applies qualitative thinking in framing the research design, noting that qualitative research is typically not interested in cause and effect. The requirement is rather to 'discover', 'generate', 'explore', 'identify' or 'describe' events in an exploratory manner. Qualitative research questions generally start with 'how', 'why', 'in which ways', 'to what extent' or similar open-ended main questions and research sub-questions. As a guiding principle, interpretivists construct meaning by interacting with the phenomenon of study in its context and by collecting data using semi-structured, open-ended questions. Interpretivists make sense of phenomena based on data gained from the personal experience and historical and societal perceptions of interview respondents. There are many other interpretivist approaches, including transformative interpretivism, which argues that most research methods do not fit marginalized groups and individuals, a view shared by the critical theorist. A variety of studies in IS and CY, particularly studies involving process, people and institutions, will find interpretivist epistemology suitable

(continued)

Table 6.4   (continued)

	Computer science (CS) research	Information systems (IS) research	Cybersecurity (CY) research
Pragmatism	While positivism dominates research in CS, pragmatist beliefs can drive some aspects of research, accepting the fact that more than one approach can support the study. The pragmatist recognizes that no single epistemological approach is effective in research and that multiple perspectives on a problem can lead to gathering the most useful data, as a basis for the most valuable insights. For example, creation and design research may require quantitative methods to carry out experiments to develop the model, as well as qualitative methods to gather user requirements to design and develop proof of concept, prototype or artefact. In so doing, positivist and interpretivist approaches are combined in a single study	Since information systems research often engages in cross-cutting studies that span technology, process, people and organization, a positivist approach on its own may not be sufficient to respond effectively to the research problem. Information systems research is often multidisciplinary or interdisciplinary, requiring pragmatism to integrate more than one research approach within a single study. The researcher can integrate quantitative and qualitative methods in a meaningful way. For instance, a usability study in artificial intelligence may involve integrated or mixed methods to address the problem from a technology and from a people perspective	Cybersecurity is a multidisciplinary domain that incorporates a wide range of fields including technology, management, policy and regulation. Research in this field requires knowledge of the national cybersecurity landscape and of global trends in cybersecurity policy, regulation and management. Hence, research in this fields lends itself to adopting mixed methods. For instance, a study in cybersecurity risk management may combine constructivist and critical realist approaches, in order to understand the organizational reality

(continued)

Table 6.4 (continued)

	Computer science (CS) research	Information systems (IS) research	Cybersecurity (CY) research
Pragmatism is an appropriate epistemology in situations where the nature of the research problem is too complex to address from a single philosophical perspective. It focuses on practical applied research: 'what works'. It focuses on solving the problem, rather than on purist research philosophy. The researcher should carefully choose the strategies, methods, techniques, etc. that would be suitable for data collection and data analysis			
Realism	In computer science, both direct and critical realists can utilize comparative research or hypothesis testing, for example, in exploring the value of algorithms or the success of particular features of software design	In the field of information systems, researchers can apply the principles of direct and critical realism, for example, in highly focused case studies, but more typically IS research relies on pragmatism to explore comparative analysis and hypothesis testing that spans technology, process, people and organization elements, for example, in addressing research problems that relate to quality of service and quality of performance	Cybersecurity research operates in a complex environment and direct or critical realism can be applied. However, the interconnectedness of the dimensions of a particular cybersecurity research problem makes it impractical to rely on realism alone. Pragmatism is more appropriate in the field of cybersecurity
Realism is regarded as metaphysical (aiming to understand the relationship between the mind and the physical universe) but can also be understood and applied in the context of epistemological thinking where researchers attempt to get as close to the 'reality' of a phenomena as is possible, given the constraints of the human mind			

6.3   Research Design

Research design aims to present a clear map of the activities required to produce knowledge, including the formulation of the research problem and hypotheses, the crafting of the analytical framework and the selection of methods for data collection, data presentation and data analysis. The research design includes an action plan of qualitative, quantitative or mixed-methods approaches and can be presented in a visual format; see the generic diagram in Figure 6.3. The researcher can adapt this diagram to their own study, filling in the specifics of research design, research philosophy and research methodology.

Figure 6.3 *Foundational
Elements of Research
Design and Methodology*

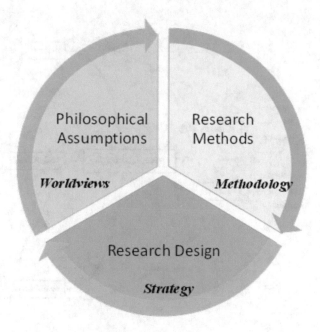

6.3.1 CS, IS and CY Research Design

The starting point for constructing a research design is understanding the environmental context and the interplay between the variables (quantitative research) or dimensions (qualitative research) of the research problem. The researcher can use the funnel strategy described in Chapter 4 of this book to develop the ideas of the study. Secondly, adopting a carefully chosen research philosophy will guide the choice of research methods, the third component of research design. Research design is necessary, in order to answer questions validly, objectively, accurately and economically. It allows for setting up conditions for collection and analysis of data in such a manner that gives relevance to the research purpose. It is an all-encompassing strategy to integrate the diverse elements of the research study in a comprehensible and rational manner. It is important to emphasize that research design is the starting point of the external evaluation of a researcher's work by examiners. The philosophical approaches and methods of enquiry as described in the research design point to whether the research will have a valid or justifiable outcome (Onwuegbuzie & Leech, 2006). In other words, the researcher must present a clear, logical and convincing argument to the examiners and readers that the research design will yield valid and reliable results.

Figure 6.4 shows five elements of research design that can guide a researcher in the fields of computer science, information systems and cybersecurity. Since the researcher must decide whether a positivist, experimental research design is needed or whether an interpretivist, qualitative research design is needed, the research design mind map can assist in guiding that decision-making process. Together

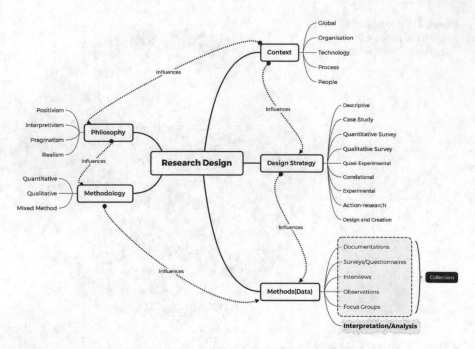

Figure 6.4 *Conceptualization of Elements of Research Design*

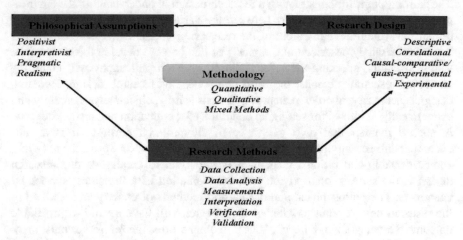

Figure 6.5 *CS, IS and CY Methodology Thinking Framework*

Figures 6.4 and 6.5 can guide the researcher in conceptualizing a workable research plan, setting out the procedures, activities and tasks required to complete the study. The goal is to present a proposal that ensures that the research design is suitable to obtain valid, objective/subjective and meaningful answers to the research questions, since research is about creating meaning from the relative chaos of ideas and data. Only sound research design can ensure that the evidence-gathering will address the research problem effectively, in an unambiguous way. Reviewers and examiners take great care in considering the quality of research design and its effectiveness in addressing the research problem.

The following steps can be used to clarify the methodology and frame the research design and methods:

Step 1: Since you have identified the research problem clearly and concisely, decide whether the research philosophy is positivism, interpretivism, pragmatism or realism. Paying attention to the research context can guide the choice of research philosophy.

Step 2: If the research philosophy is positivist, then decide what type of positivism applies, for example descriptive, correlational, causal-comparative, experimental or quasi-experimental. If the research philosophy is interpretivist, then decide which type of interpretivism applies, for example phenomenology, constructivism, social constructivism, symbolic interactionism or another type of interpretivism. Explore available material on interpretivism to enable a deeper understanding of this aspect of research design.

Step 3: Selecting a research philosophy will help to clarify the research methodology: If you chose a positivist research philosophy, then the methodology will be quantitative or mixed methods. If you chose an interpretivist research philosophy, then the methodology will be qualitative. If you chose a pragmatist research philosophy or a realist philosophy, then the methodology will be qualitative or mixed methods.

Step 4: The next step is to clarify the detail of the research methods, deciding whether to adopt a quantitative survey, a design and creation method, a qualitative survey, a case study or action research.

Step 5: With respect to data collection methods, descriptive research may adopt a quantitative survey design, while a constructivist case study design could include interviews, focus groups, documentary review and observation (or some of these). For further guidance on questionnaires, interviews and focus groups, read Williamson (2018).

Step 6: The final step in research design will include deciding on the methods for evaluating the design (Oates, 2006) prior to submission to critical readers/assessors. At this juncture, the researcher should appreciate that the research problem determines the nature of the research design, not vice versa. The understanding and knowledge of the problem is central to the choice of the approaches to data collection and data analysis.

Next, we look at some types of research design that are useful for CS, IS and CY studies.

6.3.2   Types of Research Design

This section discusses a limited number of types of research design, focusing on those types commonly used in the fields of computer science, information systems and cybersecurity research. The postgraduate researcher should explore research design more broadly. A useful website for the novice researcher is *Scribbr*, see for example, the post on research design by McCombes and Bhandari (2021–2022), which we read in preparing this chapter. Here you can explore more detailed information on descriptive, correlational, experimental and quasi-experimental, cross-sectional, longitudinal, case study, exploratory and explanatory research design.

6.3.2.1   Descriptive Research Design

Descriptive research design addresses research problems that seek to understand who, what, when, where and how, the relatively basic questions that enable us to describe the nature and characteristics of a particular phenomenon. This type of research design can enable researchers to obtain information about the status of the phenomenon, providing data about particular variables or circumstances. However, it cannot provide an understanding of the reasons why the phenomenon exists. It is often used as an initial step towards further quantitative research, as it can provide valuable indicators as to what variables are important for hypothesis testing. Examples in the field of cybersecurity are the description of the attitudes of organizations to global cybercrimes or the insider factor in cybersecurity breaches. An example in computer science is the description of how processors influence the performance of computers.

6.3.2.2   Experimental Design

Experimental design is an approach that allows the researcher to take control of all variables that could affect the outcome of an experiment. It uses a statistical approach to establish the 'cause-and-effect' relationship between a group of variable factors in a study. This design is suitable for research into an underlying relationship to study the likelihood that the independent variable(s) will always affect the dependent variable(s) in the same ways, attesting to the degree of the relationship. This is often referred to as a true experiment, but it is not necessarily a laboratory exercise. A true experiment is any research study where an attempt is made to control the independent variable(s) while observing the effects on dependent variables, outside of an organizational context. In real-world experiments, the measurements must be controlled, randomized and manipulated. Some examples of experimental study may include the following:

- The effect of malware attacks on individuals.
- The effect of artificial intelligence on supporting customers in a retailing shop.

- It can be used to substantiate algorithm theories, network theories, memory performance assumptions, AI theories, cyber risk theories, user experience theories, etc.

Experimental design has the following advantages:

- It helps the researcher to set the limits of the experiment by scoping the boundaries of the study using independent variables.
- The researcher is in full control of the independent variables, which allows the researcher to address the question, 'what causes an event to occur?'
- It helps the researcher to determine the dependent variables based on the independent variables.
- This design allows the researcher to determine cause-and-effect relationships between variable factors and to differentiate dummy effects from actual effects.
- It enables the researcher to determine what has happened (deductive analysis) or predict what may happen (predictive analytics).

In experimental research, the researcher is testing a hypothesis and can manipulate independent variables to measure the effect on dependent variables. The results can be quantified, calculated and analysed as the basis to prove or disprove the hypothesis.

6.3.2.3   Comparative Study Design

Comparative study design is applied to determine the relationship between two or more objects, or ideas or parameters by observing various aspects and comparing them analytically. A basic comparative design could test of which solution is better than another. For instance, two algorithms for solving the same problem can be assessed comparatively, based on certain attributes, to determine which has optimal performance. In the context of cybersecurity, a researcher can study different types of anti-virus software, based on predefined criteria, to determine their performance and rank them appropriately. In conducting this type of study, research designs such as experiment, observational study or literature survey can be the basis to generate or collect data, before applying the comparative techniques. An example is the study of the effect of accelerated computational skills training for young college students versus conventional learning.

6.3.2.4   Causal-Comparative/Quasi-Experimental Design

Causal-comparative/quasi-experimental design aims to determine the 'cause-and-effect' relationship among variable factors. In a true experiment, the researcher can randomly allocate themes to experimental and control groups. In contrast to a true experiment, a quasi-experiment is not based on random assignment of research participants to an experimental group and a control group but on assignment of participants to experimental and control groups using non-random criteria. For example, banking customers who have previously experienced cybercrime may be assigned to an experimental group and those who have not may be assigned to a control group. Another approach to quasi-experimental research is to study pre-existing groups with dissimilar experiences. Quasi-experimental design is usually used in research where true experiments may not be applied due to ethical or practical constraints. For example, the effect of malware (the independent variable) on a system (the dependent variable) can be measured through a 'honeypot' system as opposed to a live productive system. In IS and CY studies, it can be applied to observe people's behaviour in a usability study of a system, especially when investigating psychological factors between groups of users.

6.3.2.5 Correlational Research Design

Correlational research design aims to ascertain the level of a relationship between two or more closely related variable factors based on statistical data. This type of study is more interested in trends, patterns and relationships observed in the data, and the interpretation of the interconnections, than in determining causes for these patterns. In this kind of study, variables are not usually operated or controlled but simply identified and investigated in their natural forms.

Examples of correlational design include the following:

- The relationship between insider threats and cybersecurity breaches
- The relationship between the level of education and users' ease of use of information systems
- The relationship between the amount of data available and artificial intelligence performance

In all these cases, statistical analysis techniques are used to calculate the relationship between the identified variables, and often a correlation coefficient governs the association between the variables, where the value lies between -1 and $+1$ and where the coefficient of $+1$ depicts a positive relationship between the variables and -1 indicates a negative relationship between the variables.

6.3.3 Requirements Engineering

Requirements engineering seeks to design and assess the kind of solution that can solve the identified problem and can be applied in CS, IS and CY studies. This approach investigates the viability of a solution based on the analysis of relevant factors such as technical, operational and economic factors. The technical perspective is about the appropriateness of the technological application, the operational perspective is interested in how the solution will work in the real world, while the economic perspective is concerned with the costs versus the benefits. A feasibility study is the component of requirements engineering that investigates the operational perspective. Figure 6.6 shows the typical requirements engineering research design illustrating the research methods to be considered, commencing with requirements gathering, then feasibility study, then requirements analysis, leading to requirements specification.

For a requirements engineering design, both quantitative (survey questionnaire) and qualitative (interview, document review) methods may be applied individually or in combination as a mixed-methods study. For example, in CY studies, qualitative methods can be used to assess users' perspective on a suite of security controls, while quantitative methods can be used to measure the number of users who would choose a particular security control, in order to establish the attractiveness or ease of use of such security control.

6.3.3.1   Observational Study

Observational study is applied to understand a phenomenon through observations. This involves systematic collection of data obtained from direct observation of related events to deepen the understanding either by observation and participation or by using interviews or questionnaires. This research can be conducted by observing a series of cases, and it is usually qualitative in approach. Observational study is often used in action research enabling researchers to recognize a problem or flaws in the specified organizational environment, as the basis to develop a plan to address the problem. In executing the plan, the researcher observes what materializes, reflects on the results, reviews the plan, implements again, analyses, revises and repeats the process, until the problem is thoroughly understood in its fullest possible context and resolved. For instance, a researcher is interested in understanding the threat of an insider attack and therefore creates a plan to exploit a variety of vulnerability opportunities, to observe what happens, reflecting on the outcomes and revising the plan over and over to determine several flaws that can be exploited by an insider threat.

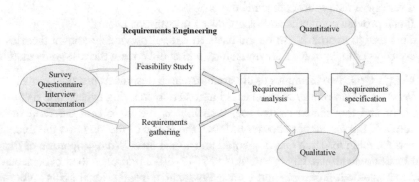

Figure 6.6   *Typical Requirements Engineering Research Design and Methods*

6.3.3.2 Simulation Studies

Simulation studies are a working example of the real problem, or the real solution, which are created and studied to explore the characteristics of the problem or solution. Simulation modelling has become the popular tool for synthesizing data obtained from multiple sources, such as global cyber-attack databases, to build a dynamic model that can provide a holistic view of attack patterns. Digitally enabled simulation, based on mathematical constructs, can be the basis to analyse the activities or the results of a physical system. A digital twin can constitute a form of simulation, but not all simulations are digital twins, since there are other types of simulations.

6.3.3.3 Case Study Design

Case study design provides the opportunity for an in-depth examination of a particular phenomenon and is suitable where a researcher would like to gain an in-depth knowledge in a particular setting, for example, in certain types of organizations, such as banks or financial services organizations. The case study could explore a limited number of banks to discover the vital characteristics, meanings and inferences about the case, which may then be generalized to the banking or financial services sector. Case studies are useful because the researcher can narrow a particularly complex research problem to a feasible easily researchable sample of cases, in order to understand the problem, as the foundation for solutions design, for practice-oriented analysis or for building theory. Case studies can use quantitative methods, qualitative methods or mixed methods for studying how a particular hypothesis, model or theory is applicable to a real-world problem. The case study has several advantages including the following:

- It provides an understanding of a multifaceted problem by thorough analysis of a selected set of dimensions (qualitative methods) or relationships (quantitative methods).
- A researcher can adopt a pragmatic paradigm to apply mixed methods and utilize a variety of data sources to study the problem.
- It can provide comprehensive accounts of uncommon events.
- This design approach can be the basis to extend previously known theories or concepts or to reinforce a previous study which did not use the case study method.

Next, we use two examples to illustrate research designs in practical terms.

Where a cybersecurity researcher is interested in studying cybercrime risk management and the related skills base, then that is the topic (not the title) of the research. If the researcher chooses to focus the scope of the study on the financial sector, the main research question could be framed thus: What proportion of financial institutions affected by cybercrime in 2018–2022 improved their cybersecurity risk management processes and their cybersecurity professional skills? Notably, the phrase in the research question 'what proportion' indicates a positivist epistemology and by implication a quantitative inquiry that may require statistical inferences. Also, the question indicates the context and timeframe of the study. This research design can be presented in a diagrammatic way. With this in mind, Figure 6.7 depicts a quantitative descriptive research design framework, showing the various interlinked elements. The context is organization, process and people, which are dependent variables of cybercrime risk management. Based on the main research question, the research operates in the quantitative paradigm, and the researcher can adopt a descriptive and deductive approach.

Designing electronic voting systems has a CS, a IS and a CY component. However, from a computer science perspective, a researcher may be interested in determining how algorithms for an electronic voting system can promote a free and fair election. The study may be scoped in a way that does not include the people element, making it CS rather than IS research. Although the people element is relevant to the effectiveness of electronic voting systems, the researcher is interested in creating algorithms and software that enable free and fair elections, without being affected by stakeholder bias. Figure 6.8 depicts a research process based on a design and creation methodology. This research can be categorized as 'learning via making', which combines the philosophical approach used (ontology, epistemology, axiology) and established scientific or engineering principles of software development to frame the study (Oates, 2006). Awareness of the context, the research strategy, methodology and data collection and data analysis methods, as well as other activities relating to the study, is necessary to construct the research design.

As can be deduced from Figure 6.8, the study must include a theoretical and conceptual understanding of the philosophy of e-voting systems and an understanding of the software development life cycle (SDLC), as these are vital components of the study. These ideas will influence the researcher in the way she/he thinks about designing the applications that will address the problem. Applying deductive

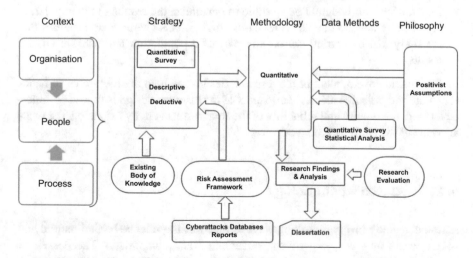

Figure 6.7 *Quantitative Descriptive Research Design*

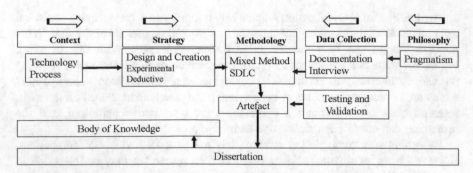

Figure 6.8 *Design and Creation Research Design*

reasoning to experimental design, the researcher can identify and draw up the arte-fact requirements and specifications. Conversely, the outcome of the research and the artefact should be tested and evaluated based on the conceptualized require-ments and specifications to ascertain that the product has satisfactorily addressed the problem.

In summation, effective research design provides the researcher, the supervisor, the reviewers, and the examiners with the assurance that the study is based on a feasible plan and has a high degree of validity. Effective research design should be characterized by the following:

- Objectivity: The assumptions, concepts and ideas should limit bias and promote neutrality.
- Reliability: The design should ensure repeatability; in other words, it should lead to similar results each time the approach is applied.
- Validity: The design should ensure that the study can be validated, both philo-sophically and methodologically.
- Generalisability: It should be possible to generalize the results to the population of the study beyond the particular cases studied. This characteristic implies that the study can be carried out on any part of a population and achieve similar results.

Systematic development of the research design can help the researcher to better articulate and organize ideas. If any aspect of the research design is flawed or under-developed, the quality and reliability of the final results and, by extension, the over-all value of the study will be weakened.

6.4   Research Methodology

A research study involves seeking ways to discover the science behind natural and social events by study, experiment, observation and comparison. The pursuit of knowledge through a systematic way of finding explanations for a phenomenon constitutes research methodology (Mkansi & Acheampong, 2012; Sileyew, 2019, Chapter 3). In contrast, research methods describe the specific approaches, procedures or techniques used to identify, select, process and analyse data and draw conclusions about a research problem. The broad methodology and the specific research methods influence the validity and reliability of the study. The methodology is concerned with three main issues: how the data can be collected or generated, how the data can be analysed and how the conclusions are drawn. The three main research methodologies are quantitative, qualitative and mixed methods, while there are a much wider range of research methods, as discussed below. In all these forms of research, triangulation is possible. This refers to more than one approach to data collection and also to more than one approach to data analysis. Using diverse methods to achieve triangulation enables the researcher to identify more aspects of the phenomenon than when using a single method (Bowers et al., 2013).

6.4.1 Quantitative Methodology

A quantitative approach seeks to quantify the research problem by emphasizing objective measurements using mathematical, computational, statistical and numerical analysis of data collected from documentation, surveys and questionnaires, in ways that ensure that the study has attributes of repeatability, transparency and credibility. When interviews are used, they are restricted to questions that support measurement, not reasons. The quantitative approach can be defined as strong empiricism, which is contingent on control and explanation of the problem, relying on numeric data or figures and on collection methods that are structured and systematic, and in some cases take a multistage approach. Hypothesis design and testing, by mathematical, computational and statistical means, is ideal for drawing quantifiable inferences through deductive reasoning and statistical analysis. Consequently, the quantitative paradigm requires the understanding of numeric data including interval or ratio and percentages and employing graphs or diagrams to explain the results (Creswell, 2014). Quantitative research is particularly concerned with measuring quantity; hence, we frame quantitative research questions using words such as *how many? how frequently? what proportion?* or *to what extent?*

6.4.1.1 Data Collection

Data collection in a quantitative study requires that the data has one of two basic forms either discrete or continuous, where discrete data represents finite numbers and values that are constant, and data values on a continuum have the likelihood of having fractions or decimals. For instance, if research is carried out to find out the number of security breaches that occurred in financial institutions, then the resultant data would be whole numbers, whereas research related to the investigation of physical measurements such as height, weight, age or distance can give continuous data. Quantitative data gathering techniques include experiments, probability sampling, methodical observations, longitudinal studies, surveys, polls, questionnaires, interviews and re-using data from statistical reports.

6.4.1.2 Data Analysis

Data analysis with respect to quantitative research uses techniques that can find evidential information, draw inferences and present the outcome in a manner that conveys understandable findings. Data analysis is the critical stage required to ensure that the outcome of the research makes sense and can provide the basis for informed decision-making. The following are popular quantitative data analysis techniques, which use deductive reasoning, noting that we will discuss data analysis methods in Chap. 7.

6.4.1.3 Trend Analysis

Trend analysis is a statistical analysis technique that provides insights into quantitative data gathered over a long period of time. This type of analytical method aids in discovering patterns embedded in events, or changes in a particular type of event over a specified period of time, as the basis to understand the change in independent variables in relation to the dependent variable.

6.4.1.4 Total Unduplicated Reach and Frequency Analysis (TURF)

Total unduplicated reach and frequency analysis (TURF) is a quantitative data analysis method that is used to measure the potential market reach of a combination of products and services. This is a valuable feedback mechanism, which can help to understand potential market reach in a particular consumer population and provides insight on how to improve product performance or consumer attractiveness.

6.4.1.5 SWOT Analysis

Strengths, weaknesses, opportunities, threats (SWOT) analysis can be used as a quantitative data analysis method that ascribes numerical values to signify strength, weaknesses, opportunities and threats pertinent to an event, product or service, as the basis to compare or predict comparative advantage. SWOT analysis can also be used as a qualitative method.

6.4.1.6 Gap Analysis

Gap analysis uses a matrix format to present quantitative data in a manner that aids to measure the variance between the expected results and the actual results. This data analysis is appropriate to measuring gaps in performance, or quality of service, and provides insight into how to address (close) the gap.

There are many software-based analytical tools that can help the researcher to interpret the data and assign meaning to the results in a manner that addresses the research question. In addition, as part of the research process, the analysis must be verified and validated, based on the parameters set out in the research design. Verification can be conducted through formal review of the data analysis, using the concepts and theories in the initial design to ascertain how they helped in answering the research question. Validation may require external evaluation to determine the correctness of the research process, findings and analysis.

6.4.2 Qualitative Approach

The qualitative research approach is concerned with unstructured, or semi-structured, non-numerical data (Mkansi & Acheampong, 2012). It seeks to understand a specific population's experiences based on the individual perceptions (or words) of members of that study population. It uses narrative, not numeric, data. Qualitative research uses inductive reasoning (Dudovskiy, 2020), where the researcher engages in obtaining an explanation from the data based on a combination of documentary evidence, observation and/or narrative (respondent-based) evidence. Qualitative research relates to exploration, attempting to offer insight into the way persons (or groups, organizations, etc.) understand the features of the social world, including the economy. Qualitative research is descriptive or interpretive, examining events that can be observed but not quantified. Qualitative researchers seek to interpret or derive a sense of the phenomena in the way that people (key informants) perceive or attach meanings to them.

6.4.2.1 Data Collection

Data collection can be conducted using documents, observation and participant observation, interviews and/or focus groups, visual essays or photo essays and other methods. For instance, a qualitative research approach can rely on semi-structured interviews to generate qualitative data by using open-ended questions and inviting each respondent to give their own version of the phenomenon of study. The respondent is encouraged to provide an in-depth response in his/her own words. The results of qualitative data collection offer deep insight into how individuals perceive their social world. Figure 6.9 shows possible data collection methods when using a qualitative research paradigm.

Figure 6.9 *Qualitative Data Collection Methods*

6.4.2.2   Data Analysis

Data analysis in qualitative research is intricate because qualitative interpretations must be derived and constructed using a variety of techniques. Such techniques include coding of qualitative data into categories and sub-categories, discerning patterns and interconnections. In qualitative data analysis, we generally commence with open coding (assigning initial codes to the data), followed by axial coding (coding to draw connections between data based on the initial codes assigned), then selective coding (choosing or prioritizing one main category which stands out as the leading category, to which other categories are related).

In summary, qualitative research seeks to understand phenomena in their natural settings, whether these be social, economic, innovation systems, environmental systems or other systems, where the data is non-numerical and based on perceptions, interpretations and/or lived experiences of key informants. Qualitative research is subjective; it does not represent a single reality.

6.4.2.3   Mixed-Methods Approach

A mixed-methods research approach is used where either quantitative or qualitative methods alone will not be sufficient to investigate the problem (Morgan, 2013), for example, in software engineering research (Di Nonum & Tamburri, 2017). It is the systematic combination of quantitative and qualitative methods in a single study. An important characteristic of mixed-methods research is that it enables triangulation of data through the application of both quantitative and qualitative methods, potentially enriching the results of the study. The quantitative and qualitative data collection and data analysis methods must complement or supplement each other.

The quantitative data collection in mixed methods provides close-ended information that measures attitudes such as rating scales of behaviours, observations, performance, etc. The analysis is usually mathematical, computational and statistics-based, using questionnaires, interviews or experimental testing of hypotheses.

The qualitative data collection in mixed methods provides open-ended information that allows the researcher to gather information through interviews, focus groups and observations, based on the perceptions of the key informants.

Five benefits of mixed methods are the following:

Expansion: Mixed methods enables the researcher to add to previous research by the application of mixed methods not previously used to understand the same/ similar research problem.

Enrichment: Applying triangulation through mixed methods enables the researcher to find the conjunction between the measurable features and the unmeasurable characteristics of the phenomenon of study, to present a rich picture of the phenomenon, thereby improving the reliability of the results. While expansion means simply adding to what is known, enrichment means creating greater depth of understanding. With any research problem, one piece of research does not reveal all the details we need to understand the problem, so in-depth research is desirable.

Complementarity: Mixed methods can be used to provide explanation and illustration. For example, quantitative methods will illustrate, through spreadsheets and graphs, the traffic on a particular broadband network at all times of the day, month and year, while qualitative methods will explain the multiple ways in which the same broadband network is used. For example, national research and education networks (NRENs) that service African universities will use traffic graphs to understand the level of demand, while using interviews to understand the value of the network to universities as users. For a researcher, this 'state of the NREN' could be a single PhD research project.

Repositioning ideas: Mixed methods can provide a way of discovering contradictions, where the results from quantitative and qualitative research contradict each other, allowing for fresh standpoints to evolve.

Methodological development: This is an approach where the researcher uses the outcome from the first methodology employed to aid the development of the second method employed.

The mixed-methods approach is suitable for conducting CS, IS and CY studies, where these studies are multidisciplinary in nature. For instance, in the creation of an artefact using requirements engineering (a type of design and creation research), a combination of quantitative and qualitative methods can be applied for requirements gathering, requirements analysis and setting requirements specifications. The quantitative methods would apply to the technical requirements, while the qualitative methods would apply to the customer requirements. A mixed-methods approach is highly relevant in measuring and understanding service quality dimensions in institutional environments.

In concluding this chapter, we emphasize the need for researchers to build and articulate their own research philosophy, research design and research methodology, specific to the research problem and main research question. It is necessary to select a research philosophy, create a research design and select the research methodology, before the researcher can adopt particular research methods and map out the detail of the data collection and data analysis processes.

Bibliography

Babbie, E. R. (2010). *The practice of social research* (12th ed.). Cengage Learning.

Bowers, B., Cohen, L., Elliot, A., Grabowski, D., Fishman, N., Sharley, S., Zimmerman, S., Horn, S., & Kemper, P. (2013). Creating and supporting a mixed methods health services research team. *Health Services Research, 48*(6.2), 2157–2180. https://doi.org/10.1111/1475-6773.12118.

Cohen, L., Manion, L., & Morrison, K. (2011). *Research methods in education*. [eBook]. Routledge.

Collins, H. (2010). *Creative research: The theory and practice of research for the creative industries*. AVA Publications. http://oro.open.ac.uk/57775/

Creswell, J. W. (2014). *Research design: Qualitative, quantitative and mixed methods approaches* (4th ed.). SAGE Publications.

Di Nonum, M., & Tamburri, D. A. (2017). Combining quantitative and qualitative studies in empirical software engineering research. In *2017 IEEE/ACM 39th International Conference on Software Engineering Companion (ICSE-C), Buenos Aires*, pp. 499–500. https://doi.org/10.1109/ICSE-C.2017.163

Dudovskiy, J. (2020). *Inductive approach (inductive reasoning). Business research methodology*. https://research-methodology.net/research-methodology/research-approach/inductive-approach-2/

Elkind, D. (2005). Response to objectivism and education. *The Educational Forum, 69*(4), 328–334. https://doi.org/10.1080/00131720508984706

Gladden, M. (2017). An axiology of information security for futuristic neuroprostheses: Upholding human values in the context of technological posthumanization. *Frontiers in Neuroscience*. https://doi.org/10.3389/fnins.2017.00605

Goddard, W., & Melville, S. (2004). *Research methodology: An introduction* (2nd ed.). Blackwell Publishing.

Goldman, A. (2001). Social routes to belief and knowledge. *The Monist, 84*(3), 346–367. https://doi.org/10.5840/monist200184314

Haig, B., & Evers, C. (2015). *Realist inquiry in social science*. SAGE Publications.

Harvey, L. (2012–2020). Symbolic interactionism. *Social research glossary*. Quality Research International. https://www.qualityresearchinternational.com/socialresearch/symbolicinteractionism.htm

Kaspersky Daily. (2021, June 28). *Ontologies in information security*. https://www.kaspersky.co.za/blog/cybersecurity-ontology/29181/

Lakoff, G. (1987). *Women, fire, and dangerous things: What categories reveal about the mind*. University of Chicago Press.

Leedy, P., & Ormrod, J. (2015). *Practical research: Planning and design* (11th ed.). Pearson Education. https://pcefet.com/common/library/books/51/2590_%5BPaul_D._Leedy,_Jeanne_Ellis_Ormrod%5D_Practical_Res(b-ok.org).pdf

McCombes, S., & Bhandari, P. (2021–2022). *Research design: A step-by-step guide with examples*. https://www.scribbr.com/methodology/research-design/

Mkansi, M. & Acheampong, E. A. (2012). Research philosophy debates and classifications: Students' dilemma. *The Electronic Journal of Business Research Methods, 10*(2), 132–140. https://uir.unisa.ac.za/handle/10500/24076

Morgan, D. L. (2013). *Integrating qualitative and quantitative methods: A pragmatic approach*. SAGE Publications.

Novikov, A. M., & Novikov, D. A. (2012). *Research methodology: From philosophy of science to research design*. CRC Press. https://www.researchgate.net/publication/274390834_Research_Methodology_From_Philosophy_of_Science_to_Research_Design

Oates, B. (2006). *Researching information systems and computing*. SAGE Publications. https://books.google.co.za/books/about/Researching_Information_Systems_and_Comp.html?id=ztrj8aph-4sC.

Onwuegbuzie, A. J., & Leech, N. L. (2006). Linking research questions to mixed methods of data analysis procedures. *The Qualitative Report, 11*(3), 474–498. www.nova.edu/ssss/QR/QR11-3/onwuegbuzie.pdf

Rasmussen, J. (1998). Constructivism and phenomenology what do they have in common, and how can they be told apart? *Cybernetics and Systems, 29*(6), 553–576. https://doi.org/10.1080/019697298125515

Sallos, M., Garcia-Perez, A., Bedford, D., & Orlando, B. (2019). Strategy and organisational cyber-security: A knowledge-problem perspective. *Journal of Intellectual Capital, 20*(4), 581–597. https://doi.org/10.1108/JIC-03-2019-0041

Saunders, M., Lewis, P., & Thornhill, A. (2019). *Research methods for business students* (8th ed.). Pearson Education.

Schwandt, T. (2015). *The SAGE dictionary of qualitative inquiry* (4th ed.). SAGE Publications.

Sileyew, K. J. (2019). Research design and methodology. In E. Abu-Taieh, A. El Mouatasim, & I. H. Al Hadid (Eds.), *Cyberspace*. IntechOpen. https://www.researchgate.net/publication/335110374_Research_Design_and_Methodology

Williamson, K. (2018). Questionnaires, individual interviews and focus group interviews. In K. Williamson & G. Johanson (Eds.), *Research methods: Information, systems, and contexts* (2nd ed., pp. 379–403). Chandos Publishing. https://doi.org/10.1016/b978-0-08-102220-7.00016-9

Chapter 7
Data Collection, Presentation and Analysis

7.1 Introduction

Some of the most important components of postgraduate research are data collection, data presentation and data analysis, because these are the foundations that the 'new knowledge' stands on. These components establish the basis on which reviewers, examiners and readers evaluate the final research report, dissertation or thesis. In this chapter, we discuss data collection, data presentation and data analysis, relevant to studies in computer science (CS), information systems (IS) and cybersecurity (CY). Data collection and data analysis techniques differ with respect to quantitative and qualitative research, hence we will discuss some of the key characteristics of each approach in this chapter.

Quantitative data is primarily used when trying to quantify aspects of a phenomenon or answer the 'what' or 'how many' aspects of a research question. It finds and presents data that is numeric. This type of data offers an avenue to find relationships (correlation, regression, etc.) between two or more variables. The data can be collected by administering a survey design using closed-ended questions or from experiments or from simulations of real-world problems. It is suitable where statistical analysis and inferences can help answer the research question. Below are instances of quantitative data collection considerations:

- When the research incorporates the statistical (how many?) element
- When frequencies are sought to explain meanings, trends or patterns
- When control of approach is needed to allow for discovery of the unknown or unexpected
- When the research problem is artificial, for instance, an experiment in a laboratory setting
- When the data explains a phenomenon using numbers and statistical analysis (or insights)

For instance, a researcher may want to know the cyber threat level in a particular sector, for example banking, or retail or government services. To identify the cyber threat level and whether it is increasing or decreasing, the researcher may count the number and frequency of all cyber-attacks over a period of a few months or over a longer timeframe. Analysing this quantitative data would enable the researcher to make inferences about the level of cyber threats. To understand which are the most potent threats, the researcher could use a survey to ask cybersecurity professionals which threats they regard as most significant and as low, medium or high risk.

Qualitative data enables the researcher to interpret the qualities or characteristics of a research problem. The data is often collected and presented in narrative form, with the use of visualization in the form of textual tables, diagrams and other images. Qualitative data is typically collected through interviews and/or focus groups, observation and document review, using semi-structured techniques and open-ended questions. Qualitative data is non-numerical and is not precise but rather exploratory and/or explanatory. It is usually presented in the form of text that describes, interprets or constructs patterns and meaning, using coding to organize the data into meaningful categories and themes relevant to the research sub-questions. Qualitative data collection is appropriate:

- Where the study seeks to understand specific aspects of systems behaviour in its wider information systems or cybersecurity context
- Where the study seeks to understand the nature of digital transformation in an enterprise, in a set of enterprises or in a particular sector
- Where the study requires an in-depth investigation of a particular case of the phenomenon of interest

For instance, a researcher may wish to understand how to create an enabling regulatory environment for drones. They would conduct key informant interviews with knowledgeable staff of regulators, legal experts and academics, in order to collect qualitative data that shares the knowledge, experiences, lessons learnt and ideas pertinent to this research problem. Analysis of this qualitative data would provide insight into recommendations for improving drone management systems (Nkamisa, 2021).

7.2 Overview of Data Collection Techniques and Processes

To effectively generate usable data, the methods and techniques used must include the following:

7.2.1   Data Identification

This requires planning what data to collect and interrogating the reasons why and how this data relates to the research problem, the particular research sub-questions and the overall research design. At this point, it is also necessary to identify the sources of the data. When planning data collection, it is necessary to identify the population for the study and the sample from that population that will be included in the study.

7.2.2   Data Gathering

This requires the researcher to conduct primary data collection (experiments, surveys, observations, interviews, focus groups, etc.) and/or secondary data collection (document review, use of existing panel data, etc.).

7.2.3   Data Organization, Selection and Processing

This refers to the methods and techniques that enable the researcher to organize or structure the data. There are two main data types, namely, structured and unstructured data.

7.2.4   Data Presentation

Once data has been organized and understood, this data has to be written up and presented in the data chapters of the research paper, dissertation or thesis. The examiner and other readers need to be able to see the data, to construct their own meaning and to consider whether they agree, partially agree or disagree with the analysis and conclusions of the student/author.

7.2.5   Data Analysis

The data must be analysed to make deductions or draw inferences (quantitative research) or to conduct qualitative data analysis (QDA). Analysis requires interpreting and giving meaning to the data, in relation to the research problem. Creating meaning enables the researcher to build new knowledge and to present solutions to research problems.

7.3   Types of Data Collection

There are many useful sites on the Web offering guidance on types of data collection and data collection tools. One such site is the QuestionPro blog at www.question-pro.com/blog/

7.3.1   Experiments

In quantitative research, a researcher could collect data to produce and measure changes or to create difference when a variable changes. The experimental testbed instrument is ideal for determining a causal relationship between dependent and independent variables. In the research conducted for the conference paper by Cinque et al. (2014), the researchers created an experimental testbed, using software fault injection, to understand what event logs should look like when an application fails. The abstract states as follows:

Abstract – Event logs are the first place where to find useful information about application failures. Event logs are available at different system levels, such as application, middleware and operating system. In this paper we analyse the failure reporting capability of event logs collected at different levels of an industrial system in the Air Traffic Control (ATC) domain. The study is based on a data set of 3,159 failures induced in the system by means of software fault injection. Results indicate that the reporting ability of event logs collected at a given level is strongly affected by the type of failure observed at runtime. For example, even if operating system logs catch almost all application crashes, they are strongly ineffective in face of silent an erratic failures in the considered system.

7.3.2   Derived Data

This method uses data from existing data sets, usually from a variety of data sources, to create a fresh data set, by some form of transformation or extrapolation, using arithmetic formulae or aggregation.

7.3.3   Observation

A researcher can capture behavioural or activity-linked data, through observation, participant observation or through the use of sensors, or video, to monitor and record data. Internet of Things (IoT) ecosystems create a fascinating new opportunity for observation (Madaan et al., 2018); see abstract below and full text available at IEEE Explore:

Abstract – The widespread adoption of "smart devices" and sensors in various domains such as, transport, home, critical infrastructure, and wellbeing has given rise to highly dynamic data ecosystems. The data in these ecosystems is a goldmine for data-driven decision making for a variety of stakeholders. These stakeholders exploit multiple device capabilities, re-purpose and re-contextualize data collected by the devices for a number of applications. Such IoT ecosystem(s) mandate data sharing at both small and large scale. However, sensitive nature of IoT data, lack of prior knowledge of purpose of data-use, regulations and standards make data sharing a non-trivial problem. It also raises concerns of data trust, ownership, and accountability. In this paper, we scope the "IoT Observatory" infrastructure to enable various stakeholders to observe and analyse the data and methodologies in IoT enabled ecosystems. It details the architectural components of IoT observatory and describe how its socio-technical lens can identify technical and ethical challenges for sharing IoT data. The paper further explains applicability of the observatory infrastructure in a smart city ecosystem. The main contributions of the paper are: (i) the definition of trust, accountability, transparency in context of IoT observatory; (ii) identification of variables that can be observed through IoT observatory for establishing trust and transparency in data sharing; and (iii) use of data access traces on IoT observatory to explain ownership and accountability.

7.3.4   Simulations and Digital Twins

Simulation requires the researcher to build a replica of the operation, or activity, of a real-world process, to generate data using computerized simulation techniques, for example, to predict cyber-attacks, to understand malware activities, to study people behaviour and to conduct control measure reliability. It is a suitable method to determine what could happen under certain operational conditions or rules. While simulation replicates an operation or occurrence to study *what could happen*, a digital twin design replicates the actual operation or actual occurrence *that has happened* or *is happening*. Digital twin modelling enables the researcher to identify actual operational weaknesses and to build a solution and model the solution, testing whether indeed the solution solves the problem. Simulations and digital twins use relatively sophisticated experimental and observational tools and techniques.

7.3.5   Surveys

This instrument is widely used in quantitative research. Survey questionnaires can be administered online using tools like SurveyMonkey or face-to-face. Based on the chosen timeframe, the three main survey design methods are cross-sectional, longitudinal and retrospective survey designs (QuestionPro, 2020).

7.3.5.1   Cross-Sectional Surveys

The survey to collect insights from the target audience will be administered over a relatively short time interval, possibly days or weeks. Researchers rely on cross-sectional survey research methods in situations where limited descriptive analysis of a phenomenon is required. For example, we can conduct a survey about the digital adoption at a particular point in time (IS research), let's say adoption rates of mobile banking apps in 2022.

7.3.5.2   Longitudinal Surveys

This approach involves conducting the survey repeatedly, over an extended period, spread across several years (e.g. every 3 years) or possibly across decades (e.g. once every decade). We can conduct surveys on the adoption rates of mobile banking apps over the period from 2022 to 2025, building a richer and more nuanced view of mobile banking adoption.

7.3.5.3   Retrospective Survey

The survey is designed to ask respondents for historical information, spread across prior years, for example, a survey of mobile banking consumers' experience in the period of the Covid-19 pandemic 2020–2022.

7.3.6   Interviews

Usually, the researcher coordinates the process of engagement with an interview respondent and presents questions for the interviewee to respond to. This is the appropriate method where there is a requirement to collect detailed information based on individual insights, views, thoughts, ideas, knowledge, experiences and feelings. Interviews and focus groups are relevant types of data collection for qualitative research, particularly for in-depth case study research.

7.3.7   Focus Group Discussion

A focus group can be relatively small but should have a minimum of six participants, and possibly an upper limit of 20 participants, in order to ensure that all participants have the opportunity to make a contribution. Focus group interviews can be useful to invite insights from a demographically diverse group of people, whose reactions or views are studied in order to ascertain common views and differing (or widely differing) views. Focus group interviews can therefore generate a wide range of data in a relatively short space of time.

An example of the use of focus group interviews can be found in the study titled 'The lived cost of communications: Experiencing the lived cost of mobile communications in low and very low income households in urban South Africa in 2014', available   at   http://www.r2k.org.za/wp-content/uploads/R2K-lived-cost-communications.pdf

EXECUTIVE SUMMARY

In considering the lived cost of communications in urban South Africa, and noting the effective mobile substitution of voice and the emergence of mobile apps and mobile Internet, this exploratory research sought to understand how mobile phone users experience ownership of mobile phones and the cost of ownership, access and usage. The report represents an initial exploration into this "lived experience", as a basis for (i) understanding mobile communications from the perspective of the user; and (ii) a future investigation into the strengths and weaknesses of existing policy and regulation from the perspective of the lived experience.

The research, conducted through focus groups composed of participants from low-income (ZAR6,400 – ZAR3,201 per month) and very low-income (ZAR3,200 per month and below) households in three cities, Johannesburg, Cape Town and Durban, revealed three major findings.

Value for money: The experience of mobile communications of the majority of participants was limited to making calls, receiving calls, sending or receiving SMS's and instant messages. Few participants living in low and very low-income households experience the wide array of communications services and mobile Internet communications that are on offer. This is largely because price levels are out of alignment with household income levels. Despite the high price for communications relative to household income, the majority of participants used private access, not public access. This indicates the limitations of both private and public access to mobile communications and the Internet for these income groups. Most focus group participants expressed the need for affordable, low-cost mobile communications, rather than "free airtime" packages, which in their design often did not meet the needs of these consumer groups.

Consumer rights, consumer protection and scams in the mobile communications environment: The focus group sessions indicate that consumers are very concerned about possible scams that have negative financial implications, and are not clear on how to manage these risks. Quality of service was of some concern, though not major concern, but should be further explored from a regulator's perspective. Consumers participating in the focus groups have explicitly expressed the need for regulation of consumer protection, hence the agenda for social regulation should include greater attention to consumer rights and consumer protection.

Lived experience of digital futures versus policy and regulation: The focus group discussions indicate that the agenda for communications policy and economic regulation should explicitly include the needs of mobile communications users with household income below ZAR6,400 per month. This is important with respect to fostering a digital future in which the majority of households can experience the array of possible benefits from mobile communications technologies, services and content.

To collect data for this study, focus groups of 15–20 participants each were held in Cape Town, Durban and Johannesburg, with diverse participants in terms of their location, gender and household income. Focus groups enabled the researchers to gain deep insight, at four events, based on responses from over 60 participants.

7.4   Designing the Data Collection Instrument: General Guidelines

Creating research instruments requires attention to several tasks. Drawing on Gideon (2012) with respect to the use of surveys in social science and the wide range of literature on data collection design, we can create data collection instruments relevant to the twenty-first century context, and we can apply what we learn about instrument design to the computing sciences and to research problems in information systems and in cybersecurity. In each case, conducting the tasks must relate to the research problem and the research design.

7.4.1   General Objectives

The researcher must clarify why the choice of a particular instrument will generate the results required to address the research problem. Will a survey be more appropriate than an experiment? Will case study interviews be more appropriate than a survey? Will participant observation be more appropriate than interviews? Will a combination of interviews and focus groups generate richer case study data than only one of these methods? What supplementary forms of data are required in addition to web crawler searches?

7.4.2   Research Data Requirement Specifications

The researcher must construct the data collection instrument in such a way that it will generate the specific types of quantitative data required (nominal, ordinal, interval or ratio; see explanation below) or the specific types of qualitative data required (individual data, institutional data, sectoral data, financial data, policy data, etc.). The data collection instrument must be specific to the hypotheses being tested or to the research questions posed. The content of the data collection instrument must be as closely related to the hypothesis and research questions as possible. For guidance on research instrument design, a useful text is Schulze (2009, Chapter 6, pp. 116–141).

In quantitative research, we can design a survey using a Likert scale (see Table 7.1), then use Cronbach's alpha to test whether the survey is reliable. The researcher must define the measurement parameters, define the evaluation

Table 7.1 *Likert Scale Measures*

Four-point scale		Five-point scale		Seven-point scale	
Value	**Labelling of quantitative value**	**Value**	**Labelling of quantitative value**	**Value**	**Labelling of quantitative value**
4	Excellent	5	Strongly agree	7	Strongly agree
3	Good	4	Agree	6	Agree
2	Average	3	Neutral	5	Somewhat agree
1	Poor	2	Disagree	4	Neither agree nor disagree
		1	Strongly disagree	3	Somewhat disagree
				2	Disagree
				1	Strongly disagree
		Value	**Labelling of quantitative value**		
		5	Very satisfied		
		4	Satisfied		
		3	Indifferent		
		2	Dissatisfied		
		1	Very dissatisfied		

parameters and define the units of measurement where applicable (e.g. download speeds in Gbps, time, number of cybersecurity breaches).

In qualitative research, questions operate at three levels: (i) the main research question, (ii) the research sub-questions and (iii) the interview questions. In the data collection instrument, we include only the interview questions, types of questions that can be answered directly by the interview respondents, for example, Why are so few cybersecurity professionals employed in your organization?

7.4.3   Preparing the Sample

The researcher must (i) specify the population for the study (e.g. banks in Nigeria and South Africa), then (ii) determine the criteria that will be used to select the sample (sample must include large- and medium-sized banks; sample must include banks engaged in extensive digital innovation and banks with relatively low digital adoption), and then (iii) choose an appropriate sample that meets these criteria. Most quantitative studies will draw sample-to-population inferences, while many qualitative studies review a few cases relevant to a larger study population. A research population could be a well-defined group of individuals, or objects, known to have similar, or common, characteristics, for example, professionals in the banking and financial services sector. The sample population will be a reasonable number of that group; in the case of quantitative research, a statistically significant sample; in qualitative studies, a sufficient sample size to support generalizing the findings to the study population.

Quantitative instruments will apply probability or non-probability sampling. Probability sampling includes simple random sampling, systematic sampling, cluster sampling, stratified random sampling, etc. Non-probability sampling is a sampling method where the researcher uses her/his personal knowledge and experience to design the sample, for example, purposive sampling where the researcher selects respondents based on their knowledge of the phenomenon and their ability to provide valuable responses to the survey or interview questions. The various non-probability sampling techniques include convenience sampling, snowball sampling, consecutive sampling, judgemental sampling and quota sampling.

7.4.4   Creating the Instrument

The focus of this step is on writing the content and structure of the specific questions, including the metrics (quantitative) or constructs (qualitative) that will be included. The researcher must determine whether questions are open-ended, closed-ended or Likert scale-type questions. The researcher must decide the general format of the instrument, including the title of the instrument, the use of sub-headings to cluster questions, the logical sequence of the questions in each sub-section and other formatting issues. The instrument can be tested for internal validity through piloting the instrument and getting feedback from survey participants or interview respondents. Such feedback can be used to revise the instrument so that the survey questions or qualitative interview questions are framed in the most appropriate way. The reason for testing is to determine that the instrument is fit for purpose and will not generate irrelevant data, before embarking on fieldwork.

7.5   Designing Data Collection Instruments: Using Measurement Scales in Quantitative Research

There are four measurement scales used in quantitative research, noting that the dimensions of the research problem determine which measurement scale to adopt. There are four measurement scales:

7.5.1   Normal Scale (Categorical Variable Scale)

Nominal scale (categorical variable scale) uses labelling to create clarity with respect to the nature of the variables being measured, rather than giving a quantitative value, for example, Occurrence of Cyberthreat A, Occurrence of Cyberthreat B and Occurrence of Cyberthreat C.

7.5.2   Ordinal Scale

Ordinal scale is designed using a number range to indicate the responses of survey participants. Respondents choose from a range, for example, 1–5 or 1–10, representing a level of agreement or satisfaction. There is no standardized scale for what is being measured. The Likert scale is an ordinal scale, used to gauge the level of agreement of respondents with a statement measuring opinions, attitudes, experiences, etc., usually in four-, five- or seven-point scales; see Table 7.1.

7.5.3   Interval Scale

Interval scale uses the techniques for nominal and ordinal data but also provides for data in which each response is a unique interval in itself. The interval is a suitable classification of variable measures such as people's age or income range. Under certain conditions, a Likert scale can approximate an interval scale, depending on what is being measured and how. To understand how this is possible, please explore further the many available guides, for example, https://stats.stackexchange.com/questions/10/under-what-conditions-should-likert-scales-be-used-as-ordinal-or-interval-data. When using the Web to explore the meaning of terminology or resources, always search for information from five or more sites and compare the information before you apply it to your data collection design. When using measurement scales for survey design, check and double-check the survey validity. You should pilot the survey and you should check the survey design with your supervisor before collecting data.

7.5.4   Ratio Scale

Ratio scale is a type of quantitative variable measurement that includes a zero measure. The ratio scale uses the same techniques as the nominal, ordinal and interval scales but adds the zero point 'true zero'. The ratio scale uses equal intervals between the values or attributes on the scale (Table 7.2).

7.6   Designing Data Collection Instruments: Qualitative Research

7.6.1   Ethics Clearance Requirements

Data collection for some quantitative research and for most qualitative research will require ethics clearance from the relevant ethics committees at the university. Universities typically have a human research ethics committee, nonmedical, and a

Table 7.2 *Ratio Scale Measures (Example 1 and Example 2)*

Value	Labelling of quantitative value/ attribute	Description of quantitative value/attribute (example 1)
0		No impact
1	Very low	Insignificant impact – any outage that causes insignificant negative impact, damage or disruption to the organization
2	Low	Little impact – any outage that causes limited negative impact, damage or disruption to the organization
3	Moderate	Important/moderate impact – any outage that causes moderate negative impact, damage or disruption to the organization and possibly to external networks or systems
4	High	High impact – any outage that causes significant negative impact, damage or disruption to the organization and possibly to external networks or systems
5	Very high	Mission-critical impact – any outage that causes major impact, damage and disruption to the organization and possibly to external networks or systems

Value	Labelling of quantitative value/ attribute	Description of quantitative value/attribute (example 2)
0		No impact
4	Low	Little impact – any outage with little impact, damage or disruption to the organization
8	Moderate	Important/moderate impact – any system that, if disrupted, would cause a moderate problem to the organization and possibly other networks or systems
12	High	Mission-critical impact – the damage or disruption to the system would cause the most impact on the organization, mission and other network and systems

human research ethics committee, medical. Where data collection involves only documents and no permission for access to those documents is required, then the researcher requests a formal ethics waiver certificate, following the relevant institutional procedure. Where the data collection involves a combination of documents and engagement with people, in any way, whether the people are the subject matter of the data collection, or whether they are key informants, a formal ethics clearance certificate must be obtained. The ethics clearance application must include the ethics clearance application form, the participant information sheet, the informed consent form, a permission letter for data collection from the relevant organization if permission is required (e.g. banks, schools and government departments must give letters of permission), permission to take photographs or video recordings and to use those images, the research proposal and any other requirements specified in the university ethics application rules and guidelines.

7.6.2   Primary Data

Primary data is unique to the demands of a particular research study at the time of data collection. In this case, the researcher can control the nature of data that needs to be collected or generated. Primary data can be collected using surveys, experiments, simulations (quantitative data) and/or interviews, focus groups, observation and other means (qualitative data). Primary data collection may be time consuming (as a general guideline, Masters level research will require 12–15 interviews of 1 hour each; PhD interviews will require 25 or more interviews of 1 hour each), but it has exceptional value to the researcher.

7.6.3   Secondary Data

Secondary data is publicly available or can be easily obtained by requesting permission from the organization that owns the data. Secondary data can be available in a variety of forms (minutes of Parliamentary Committees, Cabinet minutes, organizational reports, reports of cybersecurity incident response teams, etc.) and can be available on various platforms, often without restriction. Secondary data is sometimes used to carry out studies without having to collect primary data, for example, using existing panel data or big data and applying data analytics tools. However, in qualitative research, primary data is almost always needed to perform effective analysis and draw conclusions. It is important to be aware that the reliability and applicability of secondary data may be in doubt, requiring further verification and validation of the data, that is, it is not a quick way of doing research since the tests for validity must still be met.

7.6.4   Interview Guide

In qualitative research, the interview guide enables the researcher to enquire into the experiences and perceptions of people most closely associated with, or knowledgeable about, the phenomenon. Interview guides are particularly useful in cybersecurity research, where the aim is to gain a deep understanding of the particular cyber threat or cyber vulnerability, in its particular institutional setting, preferably from more than one organization (e.g. banks, insurance companies and retail stores). Only IT/IS practitioners, cybersecurity professionals and senior managers in the organization will be able to provide the detailed level of data needed to understand the cyber threats being investigated. The researcher must identify the most knowledgeable persons to select as respondents, from the particular organizations or sector being studied. In addition, the researcher can select respondents from among expert groups, such as academics, researchers or industry consulting groups. The interview guide should include no more than 9–10 questions, which will take approximately an hour of interview time. The questions should be arranged to focus on the dimensions of the study, as expressed in the research sub-questions. If the study explores three dimensions (and therefore three sub-questions), then the interview guide should show those three dimensions, usually as sub-headings. Each of these dimensions can have three or four interview questions, clearly directed to collecting data related to the specific dimension, adding up to 9–10 interview questions. Questions should be semi-structured, open-ended questions. Semi-structured questions implies that all respondents will be asked the same set of questions. Open-ended questions means that the way the questions are framed encourages the respondent to share their responses with a much freedom as possible, without aiming for a standard or predetermined response.

7.6.5   Focus Group Guide and Protocol

Creating focus groups is particularly useful in forms of qualitative research where the researcher seeks to gain insight into group thinking, group behaviour and perspectives of an audience for a particular market. Focus groups are also highly efficient ways of collecting a reasonable volume of data in a reasonably short period of time. In creating a focus group, the researcher seeks participants with knowledge or interests closely related to the research problem. Similarly to the interview guide, the focus group guide should have no more than 9–10 questions to be posed to the participants, clustered into the same number of dimensions (either three or four dimensions). Given that a focus group will have between 6 and 20 participants, even with 9–10 interview questions, the focus group interview can take up to 2 h, because many participants will respond to each interview question. The ethics clearance protocol for focus groups is different to individual interviews, because there is no anonymity or confidentiality in a focus group setting. Focus groups will be relatively easy to establish for information systems research but may be more difficult to convene for cybersecurity research because of the sensitivities and wider confidentiality that applies to cyber incidents.

7.6.6   Observation

In this form of data collection, the researcher is an uninvolved observer, meaning that the researcher is not an employee of the organization or is not otherwise connected with the site of study but is observing as a researcher only. In order to conduct observation, the researcher must design an observation guide, which is a set of questions that will guide the process of observation and draw the researcher's attention to particular items or activities or processes that should be observed, in order to record relevant data. The observation guide is not dissimilar from an interview guide; it should pose questions to the researcher about what she/he is observing.

7.6.7   Participant Observation

In participant observation, the same applies as discussed under observation, except that here the researcher is a participant in some way. The researcher may be a staff member of the organization being studied or may be a manager attending the meetings in which observation is taking place. The researcher may be a bank employee, or local government employee, studying enterprise resource planning (ERP) systems. The researcher as participant observer must seek permission from the organization to collect data and must create an observation guide to direct her/his attention to what is being observed.

7.6.8   Document Review

Most qualitative research can use documents to extract secondary data that will supplement and complement the data collected through interviews, focus groups or observation. Appropriate documents include internal organizational documents such as guides or manuals, annual reports of organizations, risk management reports, reports to the Board of Directors, industry standards, public policy documents, legislation, published decisions of industry regulators, sector regulators and documents of cybersecurity incident response teams (CSIRTs). Where these documents are in the public domain, no permission is required to use and cite them. Where these documents are not in the public domain, permission must be sought from the institution to use and cite them.

7.7   Data Organization, Selection and Processing

7.7.1   Structured Data

This is data which has been organized, formatted or standardized. Creating structured datasets makes it easier for the researcher to input, search and manipulate data relatively quickly.

7.7.2   Unstructured Data

In the CS, IS and CY domains, unstructured data may include data from web crawler searches, emails, file server documents, content from social media, video, audio and unstructured big data files (Mahmud et al., 2020). Researchers can extract information from unstructured data (Adnan & Akbar, 2019), organize unstructured data to create structured data (Abdullah & Ahmad, 2013) and build business intelligence that can inform decision-making and action (Rao, 2003).

Relatively basic quantitative data can be structured in multi-column tables, while complex quantitative data will require the use of data modelling techniques including graphs and dashboards. Qualitative data can be processed by applying coding structures, using software tools such as NVivo 9 and Atlas.ti 9. The coding structures enable the researcher to build a coding system, distinguishing between baseline codes and dominant or super codes. Super codes cluster related data. The coding system enables the researcher to categorize the data, as the basis for analysis, noting that coding does not constitute analysis. In this coding process, the researcher must apply their mind to what data fits at category level (all the sub-categories) and what data fits at sub-category level. The researcher should also consider how many layers of sub-categories are visible in the data, whether there are only two layers (categories and sub-categories) or whether there are three layers (categories, first level sub-categories, second level sub-categories). Usually, two layers provide sufficient layering to analyse the data.

There are many online videos which offer detailed explanations of how to use the relevant software to code and process data:

Examples
Atlas.ti. (no date). Atlas.ti 9 Windows – User Manual: Building a code system. https://doc.atlasti.com/ManualWin.v9/Codes/CodeSystem.html
Atlas.ti. (2020, January 31). Creating a coding frame with Atlas.ti by Susanne Frieser.   https://atlasti.com/2020/01/31/ creating-a-coding-scheme-with-atlas-ti-by-susanne-friese/

7.8   Data Presentation and Data Visualization

The initial steps involve modelling and describing the scope of the data, organizing the data into a formal structure and coding the data in logical patterns and sequences. Data can be presented as text, figures, photos, images, tables, graphs, maps, diagrams of building blocks of a model, coding diagrams and complex visualizations of data, including data analytics designs such as heatmaps or data analysis dashboards. There are many freely available software tools, some available at no cost for the basic tools, that enable the researcher to create effective visualization of the data. In the decade of the 2020s, students in the fields of computer science, information systems and cybersecurity are encouraged to incorporate visualization into their data presentation and data analysis chapters. The aim is to visually show the relationship between diverse data sets, to stress the nature of specific features of the data or to map data geographically.

Data are usually obtained in a raw format that most readers will not understand. The raw data must be processed, organized and structured so that the researcher makes sense of it and decides how to present data to the audience (examiners and readers). In addition, the researcher must decide in which format(s) to present the data. Each format serves a different purpose. For instance, tables, graphs and charts are ideal for presenting quantitative data. Many visualizations, such as scatter plots, density maps and correlograms can be drawn with Python. Have a look at the Python Graph Gallery at https://www.python-graph-gallery.com/ to give you some ideas (Table 7.3).

7.8.1   Textual Form

Processed data can be presented with effective use of concepts, themes, categories and sub-categories of data. Good text formatting includes a maximum of three levels of headings and sub-headings, that is, Chapter 4, the chapter heading; 4.1, the first level sub-heading; and 4.1.1, a sub-section of the first level sub-heading:

Example
Chapter 4: Data on cyber threats in the banking sector in Nigeria

4.1 Frequency of ransomware attacks

   4.1.1 Frequency of ransomware attacks for the whole month
   4.1.2 Frequency of ransomware attacks on paydays

4.2 Frequency of phishing attacks on customers

7.8.2   Tabular Form

Tabular form presents data to distinguish, classify or associate a variety of datasets. Data presented in tabular formats and in Excel spreadsheets can be transposed into visual charts:

Table 7.3 *Types of Quantitative Data Presentation*

Time series data	Bar charts	Combination charts	Pie charts
Scatter plots	Extrapolation tables	Geo map	Scorecard
Bullet charts	Area chart	Text and images	Column chart
Stacked bar chart	Heat map	Correlogram	Spider diagram

Examples

A comparison table with data on the GDP of countries and the corresponding levels of mobile penetration and Internet penetration

A comparison table of common cyber-attack vectors across the 54 African countries

7.8.3   Graphical Form

A graph can reveal trends and relationships within the dataset such as variations or deviations over time, correlations and frequency distribution. Usually, the input for the graph is drawn from another type of data, for example, data organized in an Excel spreadsheet. Bar graphs, line graphs and pie charts all use tabular data as inputs.

7.8.4   Visual Charts

Visual charts make research findings quicker and easier to interpret, for example, see below a visual representation of data coded using Atlas.ti 7. The code at the centre of the image is the main category of data (sophistication of Internet usage), while the four colour coded boxes are the sub-categories of the main category (Figure 7.1).

There are a variety of entry level software tools that can assist in data presentation and analysis, of which some popular tools are Atlas.ti 9, Google Analytics, Microsoft Excel, NVivo 9 and SPSS. Choosing the right tool depends on the data organization and analysis tasks that the researcher will undertake. The following guidelines can assist in making the right choice:

- Frame the data presentation to reflect the objectives of the study and draw a list of types of data to be formatted.
- Make data comprehensible by sorting and grouping, eliminating unnecessary data and arranging in the required form.
- Produce charts and graphs to help to add visual aspects and analyse trends: Transpose the data from the raw data format to the structured data format, for example, from interview transcripts to coded data using Atlas.ti 9 and then to thematized discussion in the text of the research paper or from Excel spreadsheets to graphs.
- Data presentation should be systematic and logical, following a clear sequence of arguments or responses to the research sub-questions, to support the study objectives.
- Present only the data and initial interpretation that are necessary to answer the research sub-questions, to make the findings clearer.

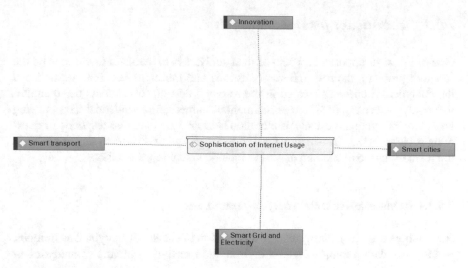

Figure 7.1 *Sophistication of Internet Use*
Source. Mosina (2020).

7.9   Data Analysis

Data analysis aims to create meaning from the data, as the basis for providing the concluding response to the research sub-questions and therefore to the main research question. The analysis of survey, experimental and simulation data can include descriptive, inferential, correlational and predictive analytical techniques. The analysis of data from interviews, focus groups, observation, document review and other qualitative data can lead to interpretation; model/framework design; recommendations for policy, law, regulation and practice; and theory building.

Data presentation and analysis are closely related activities, noting that the analysis that the researcher makes must be based on the evidence contained in the data. Data analysis may involve deductive reasoning, inductive reasoning and/or counter-inductive reasoning.

To conduct data analysis, the following are crucial to achieve credible results:

- Choice and application of analytical thinking approach
- Choice and application of quantitative or qualitative analytical tools or both for mixed-method studies
- Data sets that have high levels of internal and external validity
- Data sources that have high levels of reliability
- Ability to see errors in analysis
- Effective explanation of the analytical techniques used
- Effective argumentation in relation to the analysis provided

While the supervisor can provide guidance, the researcher must develop the capacity to check and see her/his own errors and read the methodology materials more carefully to understand how to correct those errors.

7.9.1   Techniques for Data Analysis

One of the most important aspects of data analysis is to maintain total focus on the research problem, the research sub-questions and the main research question and the analytical framework created in the research design. Researchers use quantitative analysis techniques where experimental, numerical or statistical data is being analysed and qualitative analysis techniques to analyse data that responds to questions such as 'why,' 'in which ways', 'how', etc. Qualitative analysis is usually in the form of narrative text and may also include visual representations.

7.9.1.1   Quantitative Data Analysis Techniques

Researchers engage in the systematic application of statistical analytical techniques, or they use data mining techniques, simulation analytics, predictive analytics or other techniques.

Techniques based on mathematics and statistics include descriptive analysis (historical data, key performance indicators), dispersion analysis (dispersion in the area onto which a data set is spread), regression analysis (modelling the relationship between a dependent variable and one or more independent variables), factor analysis (relationship between a set of variables), discriminant analysis (a classification technique in data mining that identifies what makes two groups different from one another) and time series analysis (measurements spanning across time).

Techniques based on artificial intelligence and machine learning include artificial neural networks (ANN), decision trees, fuzzy logic, deep learning and natural language processing.

Techniques based on visualization and graphs include column chart, bar chart, line chart, area chart, pie chart, funnel chart, word cloud chart, Gantt chart, radar chart, scatter diagram, bubble chart, gauge, frame diagram and mind map. They can be drawn with many dynamic software tools, for example, Visio, XMind, SPSS and Minitab. It is important to remember that creating the visualization does not constitute analysis. It lays the foundation for the researcher to make inferences because we can see the data presented in a structured and relational way.

Let us look at three examples of techniques used in quantitative data analysis:

Data mining is used extensively by experts to extract meaningful information from the huge databases that are generated in the course of business, for example, data from international companies or public health data. This is often the starting point in decision-based research. Big data is typically stored in databases called data warehouses or data marts. Data mining requires a student to use statistical tools to explore the data for interesting relationships that can be analysed, such as finding a gap in a use case scenario in online shopping or pinpointing areas of increasing demand in public healthcare. This involves pattern discovery and the prediction of trends and behaviours. Data visualization techniques can guide the researcher to gain a clearer understanding of the data in formats such as clusters, networks or tree models or the arrangement of a set of classifications (for instance, the attributes of valid access points in a hotspot).

Meta-analysis (Van de Kaa et al., 2007) deals with quantitative analytics used to systematically assess the results of previous research studies in order to derive new conclusions about those studies. We may use meta-analysis in cybersecurity studies to understand trends and patterns; see, for example, the publication Assessing cybersecurity: A meta-analysis of threats, trends and responses to cyber-attacks https://www.researchgate.net/publication/319677972_Assessing_ Cyber_Security_A_Meta-analysis_of_Threats_Trends_and_   Responses_to_ Cyber_Attacks

Important steps in meta-analysis include the following:

- Definition of the issues to be investigated.
- Identifying previous published or unpublished research results that have valu-able data on the defined issues. It is important to choose studies with high-quality data so that your foundation for analysis is sound in terms of the validity and reliability of the initial studies that you are relying on.
- Using the data from the identified studies, find common variables that can be used to detect significant relationships.
- Decide on the purpose of the meta-analysis, for example, to track a particular variable across all the studies and present the results to indicate its importance.
- Carry out the statistical analysis to compare or compute significance levels.
- Report the results and discuss the limitations of the research and recommend further research in the subject area.

The researcher will encounter several challenges in conducting meta-analysis, including the following: (i) a wide range of methods and statistical techniques would have been used in previous studies, making comparison and combination difficult to explain and motivate; and/or (ii) published works may only record successful outcomes where statistically significant results were achieved, leaving the other test results unrecorded. This can lead to an over-optimistic result in the meta-analysis. The researcher must consider what measures can be used to com-pensate for these potential weaknesses in applying meta-analysis.

Predictive analytics using simulation models uses easy-to-use tools to manage data and address research questions; see explanation and video on Pfizer's virtual lab – pharmaceutical analytics and simulation available at https://www.anylogic. com/blog/predictive-analytics-using-simulation-models/

Researchers should explore the extensive material including video material available online as this emerging field of predictive analytics will be exceptionally valuable where applied to problems in agriculture, in manufacturing, in health ser-vices and in other key sectors in the African economy.

The next step is to evaluate the implications of the data analysis, as the basis for deriving valid conclusions. Analytical procedures enable a researcher to eliminate unnecessary details or redundancy in the data in order to provide the basis for draw-ing inferences (Rajanikanth & Kanth, 2017). At the data analysis phase of a study, analysis is a continuous and iterative process.

7.9.1.2   Qualitative Data Analysis

As previously stated, it is important to start the process of qualitative data analysis by revisiting the original research design, including the main research problem and its three or four (not more than five) dimensions, the main research question and sub-questions (not less than two, not more than five) and the analytical framework diagram that shows the main problem and its dimensions.

Keeping in mind your own research problem and its dimensions, the key steps in qualitative data analysis include (Leedy & Ormrod, 2015) the following:

Step 1: Build on the initial process of open coding, the process of organizing and labelling the data into categories and sub-categories discussed in Sect. 7.6, by writing up a narrative analytical discussion of the meaning of the data in each category and sub-category, using inductive reasoning. Write and structure the narrative systematically, relating the structure to the three dimensions of the research problem, in the same sequence as in the initial research problem and research sub-questions. Keeping the sequence is important for the examiner and other readers to follow your analytical process with ease.

Step 2: Next, conduct the process of axial coding, carefully considering which categories link to which other categories and why. It is then possible to write up a narrative analytical discussion of the nature of the interlinkages, seeking and providing explanations for these interlinkages.

7.9.2   Data Analysis Tools

Several data analysis tools with embedded functions are available to use for CS, IS and CY research. The selection of tools will depend on the type of analysis to be performed and the type of data to be analysed. Excel, R, Tableau, Power BI, Fine Report and Python are a few such tools. Research students must develop their skills in using software tools for basic and complex analysis in CS, IS and CY research studies.

7.9.3   Artificial Intelligence (AI) Data Collection, Presentation and Analysis Platforms

There are a few leading industry AI tools/platforms for CS, IS and CY research, not limited to those listed below:

- Amazon Machine Learning https://aws.amazon.com/machine-learning/ is a managed service for building machine learning models and generating predictions in CS, IS and CY research. It includes an automatic data transformation tool that simplifies the machine learning tool for novice researchers. Amazon SageMaker https://aws.amazon.com/sagemaker/ is another fully managed platform that makes it easy for developers and data scientists to utilize machine learning models (Mahajan & Naik, 2019).

- Google TensorFlow https://www.tensorflow.org/ is an open-source software library for dataflow programming. It is a machine learning framework that offers extremely easy visualization of neural networks and other useful features for researchers in CS, IS and CY.
- The IBM Watson Machine Learning Cloud Service https://www.ibm.com/za-en/cloud/watson-studio enables researchers to put machine learning and deep learning models into productive use, including performing training and scoring, which are key machine learning operations for complex data analytics; see, for example, Alsheref and Fattoh (2020).
- Microsoft Azure machine learning https://azure.microsoft.com/en-us/services/machine-learning/#product-overviewis a cloud-based tool that supports the creation, building, training and deployment of AI models. Researchers can explore this tool to achieve credible research validations.
- Open Neural Networks (OpenNN) Library https://www.opennn.net/ is a software library for machine learning and supports data analysis, with applications in many industries, for example, energy, and fields, for example, marketing.

7.10   Chapter Summary

There are various data collection, presentation and analysis tools/platforms in the fields of CS, IS and CY. For complex research problems at PhD level, the researcher may need to consider the various trade-offs to determine the most appropriate data collection, analysis and presentation approach that meets research needs. A good choice of techniques and tools will allow for effective data analysis and validation of research outcomes, especially for real-life industry projects.

Bibliography

Abdullah, M. F., & Ahmad, K. (2013). The mapping process of unstructured data to structured data. Proceedings of the *2013 International Conference on Research and Innovation in Information Systems (ICRIIS), Malaysia,* 151–155. https://doi.org/10.1109/ICRIIS.2013.6716700

Adnan, K., & Akbar, R. (2019). An analytical study of information extraction from unstructured and multidimensional big data. *Journal of Big Data, 6,* 91. https://doi.org/10.1186/s40537-019-0254-8

Alsheref, F. K., & Fattoh, I. E. (2020). Medical text annotation tool based on IBM Watson Platform. Proceedings of the *2020 6th international conference on advanced computing and communication systems (ICACCS), India,* 1312–1316. https://doi.org/10.1109/ICACCS48705.2020.9074309

Cinque, M., Cotroneo, D., Della Corte, R., & Pecchia, A. (2014). What logs should you look at when an application fails? Insights from an industrial case study. Proceedings of the *2014 44th Annual IEEE/IFIP International Conference on Dependable Systems and Networks, USA,* 690–695. https://doi.org/10.1109/DSN.2014.69

Gideon, L. (Ed.). (2012). *Handbook of survey methodology for the social sciences.* Springer.

Leedy, P., & Ormrod, J. (2015). *Practical research planning and design* (12th ed.). Pearson Education.

Madaan, A., Wang, X., Hall, W., & Tiropanis, T. (2018). Observing data in IoT worlds: What and how to observe? In *Living in the Internet of Things: Cybersecurity of the IoT – 2018* (pp. 1–7). https://doi.org/10.1049/cp.2018.0032

Mahajan, P., & Naik, C. (2019). Development of integrated IoT and machine learning based data collection and analysis system for the effective prediction of agricultural residue/biomass availability to regenerate clean energy. Proceedings of the *2019 9th International Conference on Emerging Trends in Engineering and Technology – Signal and Information Processing (ICETET-SIP-19), India*, 1–5. https://doi.org/10.1109/ICETET-SIP-1946815.2019.9092156.

Mahmud, M. S., Huang, J. Z., Salloum, S., Emara, T. Z., & Sadatdiynov, K. (2020). A survey of data partitioning and sampling methods to support big data analysis. *Big Data Mining and Analytics, 3*(2), 85–101. https://doi.org/10.26599/BDMA.2019.9020015

Miswar, S., & Kurniawan, N. B. (2018). A systematic literature review on survey data collection system. Proceedings of the *2018 International Conference on Information Technology Systems and Innovation (ICITSI), Indonesia*, 177–181. https://doi.org/10.1109/ICITSI.2018.8696036

Mosina, C. (2020). *Understanding the diffusion of the internet: Redesigning the global diffusion of the internet framework* (Research report, Master of Arts in ICT Policy and Regulation). LINK Centre, University of the Witwatersrand. https://hdl.handle.net/10539/30723

Nkamisa, S. (2021). *Investigating the integration of drone management systems to create an enabling remote piloted aircraft regulatory environment in South Africa* (Research report, Master of Arts in ICT Policy and Regulation). LINK Centre, University of the Witwatersrand. https://hdl.handle.net/10539/33883

QuestionPro. (2020). *Survey research: Definition, examples and methods*. https://www.question-pro.com/article/survey-research.html

Rajanikanth, J. & Kanth, T. V. R. (2017). An explorive data analysis on Bangalore City Weather with hybrid data mining techniques using R. Proceedings of the *2017 International Conference on Current Trends in Computer, Electrical, Electronics and Communication (CTCEEC), India*, 1121-1125. https://doi/10.1109/CTCEEC.2017.8455008

Rao, R. (2003). From unstructured data to actionable intelligence. *IT Professional, 5*, 29–35. https://www.researchgate.net/publication/3426648_From_Unstructured_Data_to_Actionable_Intelligence

Schulze, P. (2009). Design of the research instrument. In P. Schulze (Ed.), *Balancing exploitation and exploration: Organizational antecedents and performance effects of innovation strategies* (pp. 116–141). Gabler. https://doi.org/10.1007/978-3-8349-8397-8_6

Usanov, A. (2015). *Assessing cybersecurity: A meta-analysis of threats, trends and responses to cyber attacks*. The Hague Centre for Strategic Studies. https://www.researchgate.net/publication/319677972_Assessing_Cyber_Security_A_Meta-analysis_of_Threats_Trends_and_Responses_to_Cyber_Attacks

Van de Kaa, G., De Vries, H. J., van Heck, E., & van den Ende, J. (2007). The emergence of standards: A meta-analysis. Proceedings of the *2007 40th Annual Hawaii International Conference on Systems Science (HICSS'07), USA*, 173a–173a. https://doi.org/10.1109/HICSS.2007.529

Chapter 8
Research Management: Starting, Completing and Submitting the Final Research Report, Dissertation or Thesis

8.1 Introduction

Managing the research process is a vital contributor to successful completion. Research involves a precise set of tasks and activities. This chapter focuses on the practical experiences and good practices garnered by the authors when supervising postgraduate research researchers. The chapter addresses research planning, the supervisory relationship, resource management and time management, as well as domestic issues that usually impinge on research time. The goal of research management is to complete the research process and to present the completed research report, dissertation or thesis for examination.

8.2 Research Planning

Research is a systematic practice with multiple steps, in which the steps may be iterative and interlinked in the effort to address the research question(s). Research planning should commence from the very beginning of the study, being updated and augmented as the research proceeds. The aim of the research plan is to steer the research process and design relevant actions to achieve the goal. Research planning should give a detailed overview of all activities, deliverables for each stage of the research and associated timelines. The research plan should include a clear indication of when researchers will communicate with the supervisor(s). Figure 8.1 depicts a typical research planning model, which can guide researchers in developing an effective research plan, and the elements are explained below.

© The Author(s), under exclusive license to Springer Nature Switzerland AG 2023
U. M. Mbanaso et al., *Research Techniques for Computer Science, Information Systems and Cybersecurity*, https://doi.org/10.1007/978-3-031-30031-8_8

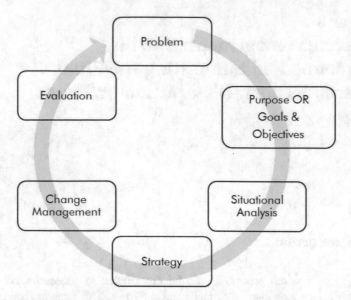

Figure 8.1 *Research Planning Model*

8.2.1 Research Problem

The hallmark of good research is to articulate and establish the research problem of a study. It is therefore important to clearly state the gap in knowledge it seeks to address and its relevance to the body of knowledge.

8.2.2 Purpose or Goal and Objectives

In some studies, we write a purpose statement. In some studies, we set goals and objectives. There is a need to articulate and establish realistic research goals and objectives with a detailed outlook of the anticipated research outcome(s). The goals and objectives should be consistent with the research problem of the study.

8.2.3 Situational Analysis

Researchers need to conduct a preliminary investigation of the research interest and area to assess whether this is a suitable area for study and a researchable topic. Furthermore, the situational analysis will help researchers identify the research gaps.

8.2.4 Strategy for Completing the Research

Effective strategy articulates not only where research is heading to and the activities required to make progress but also how a researcher will know if the progress will lead to success. The strategy should include deliverables, milestones, anticipated obstacles, challenges and workarounds. A research strategy needs to be adaptive; it needs to be responsive to the research environment. Researchers can discover that some of the underlying assumptions of the research are flawed or incomplete and need to shift perspective.

The steps below can help researchers design a good research strategy:

- Defining how to achieve the stated objectives: Describing how to achieve each intended objective is crucial in focusing on how to address the priority areas.
- Determining who is responsible: Determine who is responsible for each activity, the resources needed, timeline and deliverables. This can indicate how to reach the desired outcome.
- Evaluating (or appraising) the work: Continuous and constant reviews are important to ensure the planning performs as designed; researchers must mentally and regularly have formal reviews of the process and refine as deemed necessary.

8.2.5 Adaptability to Change in the Research Environment

Researchers must anticipate changes that may occur during the research process and address them as quickly as possible. While humans often prefer certainty because it brings clarity and expectedness, the researcher must learn to overcome fear of uncertainty and embrace the change. Researchers must appreciate that there are no areas of research immune to change, and the ability to identify the necessity for change early enough is a critical success factor for any study.

8.2.6 Continuous Evaluation

Evaluation of the progress of an enquiry is a key component of success. Researchers can adopt and adapt agile methods such as SMART (MindTools, 2018) to measure their progress. In some cases, researchers may feel that they are working hard but without getting the required outcomes. The SMART principle can assist researchers in benchmarking input and output, reflecting on all the progressions. Table 8.1 describes the SMART principle.

Table 8.1 *Description of the SMART Principle*

S	Specific	What is intended to be achieved in a specific area of focus? This requires a definition of the endpoint of a specific task or a target
M	Measurable	The task must be measurable and define a quantitative scale and timelines specifically to gauge the progress of the task or goal
A	Achievable	Based on existing knowledge, is the task achievable? What steps will the researcher take to achieve the goal or task?
R	Relevant	Is the task relevant to the overall goal, and how does the researcher know that the goal or task is achievable?
T	Timely	The task or goal must have a definite timeline and be measurable over that period. What is the expected timeframe that the task must be completed in?

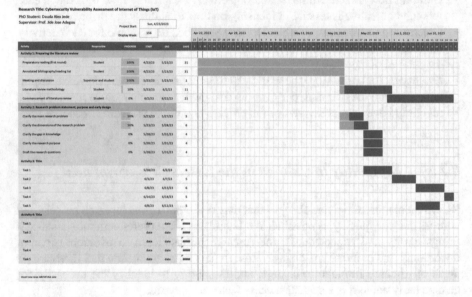

Figure 8.2 *Gantt Chart*

8.2.7   Gantt Chart

Research is a systematic and structured activity marked by several tasks and events that require proper scheduling with defined timelines. A research study can be thought of as a project using the visual Gantt chart is a visual to scheduled tasks. Figure 8.2 illustrates the Gantt chart showing its main features.

The Gantt chart can help both researcher and supervisor to keep abreast of the research process, showing start and end dates of each activity/event in one single view. It can illustrate the dependency relationships among activities or tasks as well as current schedule status. The Gantt chart can help the research work in many ways including the following:

- As a visualization and prioritization tool, it helps to keep abreast of the research progress in a single view. It shows the critical information such as the order of activities/tasks, duration, start-to-end dates, task relationship and dependencies as well as indication of progress made.

- It is helpful to improve communication with the supervisor, as a lack of effective communication between researcher and supervisor can contribute to research failures.
- It can help the researcher to carefully manage resources without overload, especially in funded studies where problems may arise when resources are over-stretched to numerous tasks and processes.
- It helps to see overlapping events and task dependencies indicating that a certain task cannot start unless the depending task is completed.
- It is an excellent tool for better time management.

The simplicity and visual effects of having all appropriate information in a single view make Gantt charts a suitable tool for effective planning of a research endeavour.

8.3 Supervisory Relationship

Researcher-supervisor relationships and expectations are crucial factors that can sustain or derail progress. A supervisor can function in a variety of roles including as a guide, as a supportive mentor and as a facilitator. At each stage of the research, it is anticipated that 'a good supervisor' attempts to support the student researcher through the rigorous processes of research design, data collection and analysis and final dissertation/thesis writing. Good supervision can lead researchers into a productive, innovative and exciting research outcome, but it is essential to note that the supervisor is only guiding the technical and theoretical work; it is the responsibility of the postgraduate students to conduct and lead their own research. In contrast, there is also the notion of 'a bad supervisor', who is either excessively supervising or, worse still, is not concerned with the research process at all. Consequently, the researcher must facilitate the supervisor to provide guidance, by providing regular drafts of the research paper. The researcher must evolve a tactical approach to make the supervisor cooperate in a positive engagement. The researcher has much more at stake than the supervisor; thus, responsibility for making the relationship mutually valuable lies more on the researcher. The following steps can help researchers develop clear expectations and mutually beneficial relationships with supervisors.

8.3.1 Gather Information and Know the Supervisor Before the Start

It is important to consider whether the proposed supervisor has the capacity to supervise the area of interest. It is also important to interact with the potential supervisor(s) to have first-hand discernment of rapport with the person(s). The notion that a scholar is an expert in a particular field or has published outstanding articles in the subject area doesn't automatically make him or her well-suited to supervise researchers. The face-to-face meeting can afford the potential researcher the opportunity to scrutinize the supervisory disposition in terms of engagement and openness to potential research interest. Paying attention to tendencies of indifference or opposition to the research interests is an indicator that should concern the researcher.

8.3.2   Understand the Expectations

It is imperative to understand the ground rules and expectations before the commencement of the research study and formal supervision. Researchers should develop an effective work plan and work ethic that clarifies the kind of guidance and assistance required from the supervisor(s). It is not usually out of place for researchers to be politely assertive when seeking advice. However, researchers should do the required work in advance, without which advice is not possible.

8.3.3   Fashion Good Communication Strategy

Developing a good communication strategy will go a long to help glue the researcher-supervisor relationship (Hyttinen, 2017). Researchers need to devise effective ways to communicate with supervisors regularly, both informally and formally. The idea is to constantly keep in touch with the supervisor through formal meetings to review the progress of work while soliciting further instructions. Evaluating which communication channel is mutually best suited for the conversation for both parties is important.

8.3.4   Consider That the Supervisor Has Other Matters to Attend To

Supervisors have busy schedules and many pressing matters that demand their attention. When not receiving timeous responses from the supervisor, always resort to gentle and courteous reminders. Never sound disappointed with the supervisor, but persistently find ways to draw his or her attention without overly showing displeasure. Researchers should always bear in mind that it is their career that is at stake, not the supervisors'. Researchers should always be prepared for meetings with relevant questions, ideas or requests. Researchers should make an effort for timely submission of work that requires the supervisor's attention, giving the supervisor a chance to review it before asking for feedback.

8.3.5   Alternative Supervisor

If despite all the planning, communications and polite reminding, the researcher and supervisor seem not to be getting along, or there are feelings of great displeasure that the relationship is unfavourable to both the parties and the study; researchers have the option to deliberate on alternative supervision chances with both the current supervisor and the postgraduate administration. Researchers should be aware that for whatever reason this concern arises, solving the issues as soon as possible is the best for keeping the research work on course. Researchers should always bear in mind that the process is entirely self-beneficial, and the supervisor's goal is to assist the researcher to be successful. But in a case where the desired guidance and assistance is not being deployed, it's the responsibility of the researcher to draw it to the attention of the relevant authorities.

8.4   Resource Management

Resource management is a key enabler of research success, similar to any type of project management (PMI, 2020). Assembling resources such as people, processes and tools to accomplish the research objective is vital. It is important to highlight the importance of soft skills such as communications and organizational and emotional intelligence to manage yourself as a researcher and to manage the interview respondents. In practice, the key points to effectively managing resources include the following:

- Researchers should make an effort to identify and understand the key resources that are required to enhance the research process. This may include human and material resources. For instance, if the supervisor is not highly knowledgeable in the subject area under study, researchers may need to identify experts in the area that are willing to be of help.
- Researchers should prioritize the need for the resources and estimate the costs if any.
- Researchers should propose schedules that can meet the research study timeline.

8.5   Time Management

Time management is a crucial element of the study (Hyttinen, 2017). It can be described as the act or method of working out consciously and the control over the time spent on a specific task or activity (Baars, 2006). In particular, developing the culture of getting more done in less time amid constricted timeframe and pressures is not trivial. Therefore, research is a rigorous and task-intensive exercise, requiring tactical use of resources and skills to work powerfully while making better use of time. Time management is a catalyst to personal effectiveness and productivity, identifying how to prioritize and schedule time for optimum impact. The full-time researcher will need to work at least 8–10 h per day including many weekends, while the part-time researcher will need to work some evenings and at least 8–10 h over weekends, preferably longer. This is because the human brain needs to be immersed in the subject matter in order to think through the many permutations of research production. Research requires dedication, and dedication requires time. Set priorities and avoid 'time-robbers'. Researchers' objectives should focus on:

- Setting priorities and managing time to meet deadlines
- Devising how to set and achieve goals
- Effectively organizing the daily tasks or activities
- Working in a team or building one
- Taking physical and mental breaks regularly

Furthermore, it is important to note that individual circumstances, such as perceived obstacles, are often the primary reason why people procrastinate, have difficulties in saying no to social engagements or struggle to delegate work which can easily be delegated. It is particularly vital in time-constraining situations when time is limited and demands are seemingly unlimited that researchers attempt to:

- Identify and prioritize required skillsets for success
- Create calmness and space for research in one's life
- Focus energy and attention on the things that matter
- Allocate time where it is most needed and most wisely spent
- Less important tasks that can be attended later can be quietly dropped

Good prioritization with careful controlling of deprioritized tasks can significantly reduce stress and help move towards a well-organized and structured process. Without such, researchers can flounder, drowning in competing demands. In simple ways, researchers can adopt the following approach:

- Prioritize all tasks based on the likely viability or benefit of the task.
- Manage time constraints, noting where the task is on the critical path of important activity and involves other parties, and the task is needed to complete to accomplish the overall objective; focus on that task. A few tools can help researchers develop a time management culture, especially paired comparison analysis or grid analysis (Bright Hub Project Management, 2011).
- Paired comparison analysis is a powerful tool and is used where decision criteria are unclear, subjective or unpredictable. It can help researchers to prioritize options by comparing each item on a list with all other items on the list. By determining in each case which of the two is most critical, the researcher can generate a schedule of priorities.
- Grid analysis can help to prioritize a list of tasks where there is a need to consider many different factors. For instance, you want to collect data, but there are many competing factors, such as library access, Internet, meeting experts, cost, quality, etc. Grid analysis can aid in taking decisions confidently and realistically.

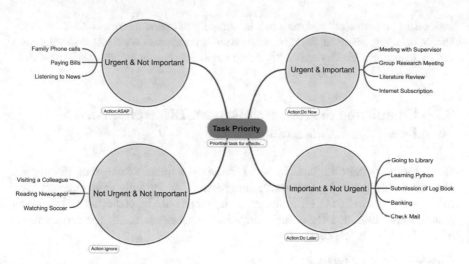

Figure 8.3 *Using Mind Map to Prioritize Tasks*

- Priority quadrant analysis, where the priority (urgent/important) matrix helps you consider whether a task is urgent or important and to assign a priority level to a task. Using this approach, researchers can ascertain whether an urgent task is important or not. In many cases, really important activities may not necessarily be urgent. The priority quadrant, which is an intersection of urgent and important, allows the researcher to prioritize which task should be done now, as soon as possible (ASAP), be later or be deleted completely. Figure 8.3 illustrates the use of the mind map to prioritize tasks using the urgent/important matrix model.

8.6   Creating Research, Home Life Balance

For any researcher, there are bound to be competing domestic needs that can make it difficult to achieve the desired outcome. The researcher should endeavour to address all domestic challenges as they are capable of derailing or completely frustrating the research efforts. A researcher with extended family, spouse and children is likely to face considerable challenges, since balancing family responsibilities with research work is not a trivial task. Consequently, it is advisable that researchers devise workable strategies to complete the research, including working late at night or early in the morning. A very helpful way to progress is for the researcher to constantly discuss the research process with the family as a way of continually reminding them that there is progress. The researcher should engage the family in interesting conversations concerning the research study and in some instances ask for their opinions or feedback on certain issues. Based on our experiences, we have found that involving family members is an effective way to get them to be compassionate about the research work and, as a result, can help to create a more conducive atmosphere at home. More so, the researcher should make consistent efforts to maintain family responsibilities by actively seeking help from family members when needed.

8.7   Completing the Research Report, Dissertation or Thesis Ready for Examination

When the researcher has completed the required number of chapters, usually six chapters for a Masters by coursework and research and usually eight chapters for a Masters by dissertation or PhD, the research defence can be presented and the completed work submitted for examination. But what does 'ready for examination' mean? Here is a quick checklist:

Checklist for Examination

Item 1: Cover page with all relevant information about the student and the institution in the format required by the university.

Item 2: All front pages, including the abstract, dedication, acknowledgements, glossary, list of abbreviations, table of figures, table of tables and table of contents. It is best to consult the APA style guide as it gives a comprehensive overview of the format and sequencing that is followed in APA style. Watch the tutorial webinar A Step-by-Step Guide for APA Style Student Papers, available at https://apa-style.apa.org/instructional-aids/tutorials-webinars.

Item 3: The relevant number of chapters is either six or eight. Here is a quick guide:

> Chapter 1 includes a very short introduction, the research problem statement, research purpose statement, research questions or hypotheses and a clearly written background to the problem.
>
> Chapter 2 includes the literature review and analytical framework, if the latter is applicable, which it is in qualitative research studies.
>
> Chapter 3 is the methodology chapter, which includes statements of ontology, epistemology, methodology and research methods.
>
> Chapter 4 is the data presentation chapter. A Masters by dissertation (MDiss) and a PhD may have three data presentation chapters.
>
> Chapter 5 is the data analysis chapter. (In the case of a MDiss or PhD, the data analysis chapter could be Chapter 7).
>
> Chapter 6 is the conclusions chapter. (In the case of an MDiss or PhD, the conclusions chapter could be Chapter 8. A few PhDs may have nine chapters, but it is good to minimize the number of chapters to eight).

Item 4: The final chapter is equally important to all the other chapters. In this chapter, the researcher demonstrates that they have answered all the research sub-questions and the main research question, that they have effectively proved or disproved the hypothesis or that they have responded effectively to the goals and objectives of the study. If the study at MDiss and PhD level required a contribution to theory, then this theoretical contribution must be clearly articulated in the final chapter, and the researcher must demonstrate and explain the nature of the contribution to theory.

Item 5: Since the researcher will have written the first three chapters many months or perhaps a year earlier in the case of a PhD, the researcher should now review the full set of chapters since a research report, dissertation or thesis is one connected whole, not just a series of chapters. The researcher must check that the content of each chapter is well aligned with the content of every other chapter. One way of checking alignment is to draw a map of all the chapters and to demonstrate to yourself how the content of each chapter is related to the content of the other chapters. In doing so, you will pick up gaps, errors and discrepancies which you can correct.

Item 6: While the researcher should use the required reference style from the commencement of the research, the researcher should now check that the reference style has been consistently applied across all chapters and in the list of references.

Item 7: The list of references comes immediately after the final chapter.

Item 8: The relevant annexures must be added, including all research instruments. The ethics clearance certificate must be added if the research uses human participants or if permission was required to use data or documents from organizations.

Item 9: The entire document must be edited for language usage, grammar and spelling, so that the research submitted for examination is clear and well written. This means that the examiner can concentrate on the quality of the research, not on the quality of expression. A well-written and well-presented research study means that the examiner will realize that the researcher has given the required attention to detail and has put in their best effort. Use Grammarly, a free online writing assistant (https://app.grammarly.com/), or another good software application, to edit your work.

Item 10: Having completed all the steps above, the supervisor can review the final paper and sign off on submission for examination.

8.8 Defence or Viva

Often called the viva, this is short for viva voce, or 'with the living voice'. The concept of defence or viva is a critical stage in postgraduate research. The postgraduate candidate must demonstrate originality and competence to the academic panel and audience in the oral defence. This is a time to demonstrate ownership of the work, proof of exceptional quality and depth of knowledge. While a formal presentation with slides will be useful, the academic panel (or supervisory committee) is less interested in the slide images than in the whether the postgraduate researcher really knows their work. Searching for information in your slides, or in the thesis document, may give the impression that you do not have a thorough acquaintance with your work.

Since the nature, structure and procedure for the defence may vary from university to university, the researcher will be guided by the postgraduate school and the supervisor(s) on the specific requirements. In most cases, the researcher will be presenting the defence or viva before an academic panel or supervisory committee, ideally composed of a combination of knowledgeable internal and external people. The defence aims to ascertain whether the work is ready for submission to examiners, that it is comprehensive and meets the requirements of scholarly merit. Thus, the researcher should be prepared to answer several questions after the presentation, implying that the researcher needs to have the capacity to confidently respond to the questions in a relaxed manner. In some universities, the defence is segmented into internal and external components, the goal being to strengthen the work through an internal process in order to guarantee success in the external defence.

8.8.1   Planning and Preparing for Defence

The first and most obvious is to know the procedure and format of the defence, the audience and mode of assessment – whether slide presentation is involved, the duration, etc. This information may be available from the postgraduate school or department, but the supervisor(s) may need to make clear some aspects of the requirements to avoid ambiguity. This is one of the crucial stages of the study, depending on the type of defence; the supervisor(s) should be involved in the planning and preparations and should approve all aspects of the report. Note, in particular, the things that can effectively aid preparation as follows:

Review current presentations: Look out for recent presentation from the postgraduate school or department, and if possible, attend several other researcher's defences before the actual defence. The aim is to learn from the experiences (ideas) of others; strengths and pitfalls and thereby reinforce the knowledge required to prepare for the defence. The researcher will observe the nuances of the presentation – how questions are raised and addressed, gaining a variety of techniques and their effects. This is helpful to examine what works or vice versa, arming the researchers to make the necessary adjustments proportionately.

Be clear on expectations: There are expectations to be met; effective planning and preparations entail mentally knowing what is expected by the defence stakeholders. The researcher should take a clue from the current presentations attended and notes taken and focus on the expectations by improving on them and anticipating what the panel or committee may look for with the kind of research.

The current state of knowledge: The researcher needs to be armed with the current state of knowledge in the literature, what is trending, the arguments and positions, prominent scholars and their disposition in current knowledge, similar ongoing research, etc. The goal is to be armed with all the related available resources influencing the topic. The researcher should focus on current articles without neglecting past literature that has gained popularity. The researcher should make note of important discourse and think critically about how the studies do not completely address the problems in the present study and how it built from what is in existence.

Articulate possible defence questions: It is important to articulate before the presentation, a list of the possible defence questions to expect and rehearse them. The researcher can do a mock presentation for the family or classmates to practise the questions and responses. If the researcher has presented at conferences, workshops or journal articles, it is most likely that some questions may be drawn from such materials. Building the level of confidence is key to a successful presentation, so practising questions and responses will go a long way to prepare the researcher.

Prepare the slides to be concise and clear: The number of slides should depend on the allotted time for the presentation; accurate timing is important to ensure the presentation of vital elements of the research such as contribution to the current state of knowledge, conclusions and future work. The use of scholarly language and prose is important to captivate the audience.

Get the report and presentation slides reviewed: It is vital that both the report and presentation slide be reviewed by a friend or a professional academic editor to provide the researchers with helpful feedback and revisions.

Practise and rehearse: The idea that practice makes perfection still holds. A week before the defence, the researcher should subject self to constant practice and rehearsal to continuously build the degree of confidence that is required to defend successfully.

8.8.2 Checklist for Defence

It is important as part of preparations to have a checklist of items to finalize the preparations. Check as follows:

- Check spelling and grammar of the report.
- Cross-check figures and tables to ensure proper labelling and referencing in the text.
- Cross-check to ensure sections, chapters, figures and tables are capitalized.
- Ensure consistent capitalization in captions.
- Verify expansion of all abbreviations at first instance.
- Avoid personal pronouns such as me, us, them, us, etc.
- Check references for capitalization of abbreviations and missing data such as page numbers.
- Know the report thoroughly and create a one-page summary of each chapter.
- Know what are the implications of the research in both theory and practice.

8.8.3   The Defence Room

The defence room remains the place to prove that the researcher is the author of the report; as such, while ensuring the mastery of the report to present the original findings that advances the frontier of knowledge, the researcher must mentally take control of the room. The researcher should appear confident and prove to the committee and supervisor(s) the capability to produce a broad-ranging and in-depth pieces of scholarly work. To this extent, the researcher must appear professional and decent comfortably and should not distract the audience by putting on an awkward or offensive (ill-dressed) costume. In some cases, the defence may be a thing of anxiety to researchers – that is, standing in front of an assemblage of accomplished scholars in the subject area and attempting to prove the work, arguments and findings can be frightening. But researchers must see defence as an opportunity to share the work and reconnoitre fresh perspectives as well as network with experts.

Consequently, the researcher must be composed, self-confident and in control of the exercise from start to the end. The following should be considered:

Consideration 1: Make the presentation within the time allocated for it, making eye contact to key stakeholders.

Consideration 2: The researcher should demonstrate the mastery of the topic far better than any or all of the scholars and other stakeholders in the room. The researcher should build extra confidence since the work is an original product that has passed through the rigour of research investigation. But the researcher should not underestimate the knowledge of the committee or panel members as they likely know the field in a much wider sense.

Consideration 3: Listen carefully to observations and questions and respond professional either in defence of the position or providing clarity. It is normal to expect to address critical questions, such as 'So what?', 'And then?' and 'What has the research accomplished that is a significant contribution to the current state of knowledge in the field?' Notably, there may be the more controversial facets of the work that may require interrogation; questions should be expected. This is an opportunity to shine by making emphasis on key areas of the research. Researchers should be bold enough in making claims for what have been achieved – the findings and conclusions. Researchers should never create doubts about the originality and value of the work. The expectation is to situate the work within the larger scheme of things, in the current state of knowledge. This is common and should be prepared for.

Consideration 4: It is allowed to seek for clarification of ambiguous questions; the researcher can always ask for the repeat of the question if so desired. It is not uncommon to enter into a dialogue with the examiners to further illuminate the researcher's position.

Consideration 5: It is common to be nervous during the defence, but remaining focused and paying attention to the audience will reduce the anxiety. Taking a moment to pause and think before answering a question is critical; when in doubt, take a sip of water to buy time and then be refocused. Provide answers with confidence and passion; most research is not about 'yes' or 'no' answer, but assertive evidence-backed answers can always change the narrative or position of the argument. For questions without readily made answers – the response should be a sort – 'the question is interesting but was never considered, it would be looked into in the future'. If there is a problem managing the fears during the defence, make go the use of visual aids to stay focused and confident.

8.9 Summary

Research management although hardly discussed in most of the research books is an important element of a research study. This chapter has provided important clues on how to manage a variety of research activities in a timely, effective and efficient manner while boosting productivity. It can help researchers treat uniquely particular aspects of the research to the continuous progress of the study and successful outcome.

Bibliography

Baars, W. (2006). *Project management handbook: Data archiving and networked services.* Version 1.1. https://www.projectmanagement-training.net/wordpress/wp-content/uploads/2015/11/book_project_management.pdf

Bright Hub Project Management. (2011, February 22). *Decision-tree vs. grid analysis.* https://www.brighthubpm.com/project-planning/107546-comparing-decision-tree-and-grid-analysis-techniques/

Hyttinen, K. (2017). *Project management handbook.* Laurea Publications. https://www.research-gate.net/publication/320101542_PROJECT_MANAGEMENT_HANDBOOK

MindTools. (2018). *SMART goals.* https://www.mindtools.com/pages/article/smart-goals.htm

Project Management Institute (PMI). (2020, January 12). *What is project management?* https://www.pmi.org/about/learn-about-pmi/what-is-project-management

Index

Printed in the United States
by Baker & Taylor Publisher Services